CUTNELL&JOHNSON

Physics 9th EDITION

Physics 9th EDITION

John D. Cutnell & Kenneth W. Johnson
Southern Illinois University at Carbondale

with contributions by

Kent D. Fisher
Columbus State University College

WILEY

John Wiley & Sons, Inc.

Founded in 1807, John Wiley & Sons, Inc. has been a valued source of knowledge and understanding for more than 200 years, helping people around the world meet their needs and fulfill their aspirations. Our company is built on a foundation of principles that include responsibility to the communities we serve and where we live and work. In 2008, we launched a Corporate Citizenship Initiative, a global effort to address the environmental, social, economic, and ethical challenges we face in our business. Among the issues we are addressing are carbon impact, paper specifications and procurement, ethical conduct within our business and among our vendors, and community and charitable support. For more information, please visit our website: www.wiley.com/go/citizenship.

To order books or for customer service, please call 1-800-CALL WILEY (225-5945).

ISBN-13 978-1-118-25437-0

Printed in the United States of America

10 9 8 7 6 5 4 3 2

CONTENTS

0 | Math Skills 1

1 | Introduction and Mathematical Concepts 19

2 | Kinematics in One Dimension 23

3 | Kinematics in Two Dimensions 27

4 | Forces and Newton's Laws of Motion 31

5 | Dynamics of Uniform Circular Motion 37

6 | Work and Energy 41

7 | Impulse and Momentum 47

8 | Rotational Kinematics 51

9 | Rotational Dynamics 57

10 | Simple Harmonic Motion and Elasticity 63

11 | Fluids 71

12 | Temperature and Heat 77

13 | The Transfer of Heat 83

14 | The Ideal Gas Law and Kinetic Theory 85

15 | Thermodynamics 89

16 | Waves and Sound 97

17 | The Principle of Linear Superposition
and Interference Phenomena 103

18 | Electric Forces and Electric Fields 107

19 | Electric Potential Energy and the Electric Potential 113

20 | Electric Circuits 119

21 | Magnetic Forces and Magnetic Fields 127

22 | Electromagnetic Induction 135

23 | Alternating Current Circuits 141

24 | Electromagnetic Waves 145

25 | The Reflection of Light: Mirrors 149

26 | The Refraction of Light:
Lenses and Optical Instruments 155

27 | Interference and the Wave Nature of Light 163

28 | Special Relativity   169

29 | Particles and Waves   173

30 | The Nature of the Atom   177

31 | Nuclear Physics and Radioactivity   183

32 | Ionizing Radiation, Nuclear Energy
     and Elementary Particles   189

   Key Data   193

What is this *WileyPLUS Companion*?

The *Physics WileyPLUS Companion* is a **quick reference** version of the main text. It is designed to accompany *WileyPLUS* online, which includes a complete digital version of the main text. This edition includes summaries of the important concepts in each chapter, as well as key tables and selected illustrations.

Does this *WileyPLUS Companion* have everything students need for class?

Yes! Besides this quick reference guide, an access code is included when purchased as part of the WileyPLUS Learning Kit, which gives students an entry to *WileyPLUS*; instructors will provide the URL where students can log on.

What do students receive with this access code to *WileyPLUS*?

WileyPLUS offers an innovative, research-based, online environment for effective teaching and learning.

WileyPLUS provides an online environment that integrates relevant resources, including the entire digital textbook, in an easy-to-navigate framework that helps students study more effectively.

- *WileyPLUS* adds structure by organizing textbook content into smaller, more manageable "chunks."
- Related media, examples, and sample practice items reinforce the learning objectives.
- Every homework problem has hints, links, and solutions to help students get the correct answer.

Chapter 04, Problem 110 GO

A mountain climber, in the process of crossing between two cliffs by a rope, pauses to rest. She weighs 515 N. As the drawing shows, she is closer to the left cliff than to the right cliff, with the result that the tensions in the left and right sides of the rope are not the same. Find the tensions in the rope to the left and to the right of the mountain climber.

65.0°   80.0°

$T_L = \boxed{884}$ N
$T_R = \boxed{814}$ N

Show Hint   GO Tutorial   Show Solution

Link to Text   Link to Text

One-on-One Engagement. With *WileyPLUS*, students receive 24/7 access to resources that promote positive learning outcomes. Students engage with related examples (in various media) and sample practice items, including:

- More worked examples
- Interactive problem-solving tutorials
- Self-quizzes
- Animations
- Simulations

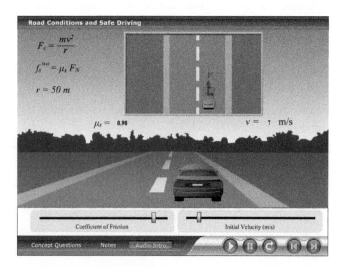

- Algebra and Trigonometry review.

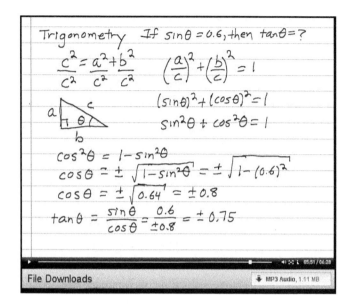

All this material can be found in the **Read Study & Practice** tab of *WileyPLUS*.

Measurable Outcomes. Throughout each study session, students can assess their progress and gain immediate feedback. *WileyPLUS* provides precise reporting of strengths and weaknesses, so that students are confident they are spending their time on the right things. With *WileyPLUS*, students always know the exact outcome of their efforts.

What do instructors receive with *WileyPLUS*?

Beyond the standard supplements—including Instructor's Solutions Manuals, Instructor's Resource Guide, Lecture PowerPoint Slides, Classroom Response Systems ("Clicker") Questions, Wiley Physics Simulations, Wiley Physics Demonstrations, Testbank, and Image Gallery—*WileyPLUS* provides reliable, customizable resources that reinforce course goals inside and outside of the classroom as well as visibility into individual student progress. Pre-created materials and activities help instructors optimize their time:

Customizable Course Plan: *WileyPLUS* comes with a pre-created Course Plan designed by the author team uniquely for this course. Simple tools make it easy to assign the course plan as is or modify it to reflect the course syllabus.

Pre-Created Activity Types Include:

- Questions for all end-of-chapter problems coded algorithmically with hints, links to text, whiteboard/show work feature, and instructor-controlled problem-solving help
- Readings and resources, including Gradable Reading Assignment Questions embedded within the online text
- Conceptual questions for each major topic in a chapter with answer specific feedback
- Algebra and Trigonometry diagnostic with links back to remedial content
- MCAT-style practice tests

Gradebook: *WileyPLUS* provides instant access to reports on trends in class performance, on student use of course materials, and on progress toward learning objectives, thereby helping to inform decisions and drive classroom discussions.

Math Skills

0.1 | Symbolic Manipulation

Algebra includes the symbolic manipulation of variables and the use of specialized techniques such as factoring, the zero-product rule, and the quadratic formula. One goal is to be able to solve an equation for one of its variables.

EXAMPLE
Solving an Equation for a Variable

If $v = 10 + 5t$, then what is t?

    This equation has two variables: v (the object's final speed) and t (the elapsed time). In this case, we are asked to determine the elapsed time. If we subtract 10 from both sides and divide both sides by 5, then the variable t will be all by itself on the left-hand side of the equation. The result is

$$t = \frac{v - 10}{5}$$

    We can also solve an equation which consists of only symbols for one of its variables. In the above equation, say that the number 5 stood for the rate at which the object was accelerating (5 meters per second squared) and 10 stood for the object's initial speed (10 meters per second). If neither of these amounts were given, then the equation would be

$$v = v_0 + at$$

    This equation has four variables: v (the object's final speed), v_0 (the object's initial speed), a (the rate at which the object was accelerating), and t (the elapsed time). If we subtract v_0 from both sides and then divide both sides by a, then the variable t will be all by itself on one side of the equation and the result will be

$$t = \frac{v - v_0}{a}$$

Therefore, as you can see, it doesn't matter whether we are dealing with numbers or symbols. The rules of algebra are always the same.   ■

0.2 | Factoring and the Zero-Product Rule

In many cases you can solve an equation by **factoring** and using the **zero-product rule:**

> If the product of any number of factors is zero, then one of the factors must be zero.

EXAMPLE
Factoring and Using the Zero-Product Rule

If $x^3 = 3x^2$, then what is x?

Your first instinct might be to divide both sides by x^2 and arrive at $x = 3$. This answer is correct, but it is not complete. The better approach is to subtract $3x^2$ from both sides, factor, and use the zero-product rule.

$$(x^2)(x - 3) = 0$$

The zero-product rule states that in this case either $x^2 = 0$ or $x - 3 = 0$. Solving these two equations yields the complete solution set of $x = 0$ or $x = 3$.   ■

0.3 | The Quadratic Formula

A special case occurs when the largest exponent of a polynomial equation is two. In this case, the equation is called a **quadratic equation** and it is commonly written in the form

$$ax^2 + bx + c = 0$$

The value or values of x can be determined by using the **quadratic formula:**

$$x = \frac{-b \pm \sqrt{b^2 - 4ac}}{2a} \tag{0.1}$$

EXAMPLE
Using the Quadratic Formula

If $5x^2 - 5x = 30$, then what is x?

If we subtract 30 from both sides and divide both sides by 5, then we have:

$$x^2 - x - 6 = 0$$

This is a quadratic equation, where $a = 1$, $b = -1$, and $c = -6$. Using the quadratic formula, we get:

$$x = \frac{-(-1) \pm \sqrt{(-1)^2 - 4(1)(-6)}}{2(1)}$$

$$x = \frac{1 \pm 5}{2} = 3 \quad \text{or} \quad -2$$

Alternatively, we could factor the equation and use the zero-product rule.

$$(x - 3)(x + 2) = 0$$

In this case, either $x - 3 = 0$ or $x + 2 = 0$, which yields $x = 3$ or $x = -2$. Notice that each method produces the same two solutions. ■

0.4 | Geometry

Geometry includes methods to determine the lengths of curved lines and the surface areas, cross-sectional areas, and volumes of three-dimensional shapes.

The length of a curved line that is a piece of a circle is found by using the **arc length formula:**

$$s = r\theta \tag{0.2}$$

where s is the length of the curved line, r is the radius of the circle of which the line is a piece, and θ is the angle measured in radians between the two radii that connect the two ends of the curved line to the center of the circle. (See Figure 0.1.)

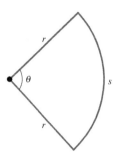

To check this relationship, we could determine the length of the entire circle, which is called the **circumference.** The angle at the center of the entire circle is 360°. If you go around the circle once, then the angle is equal to 2π radians. (In equations, radians are often written as just a number, so this would be written as 2π.)

$$s = r\theta = (r)(2\pi) = 2\pi r = C = \text{the circumference of the circle}$$

This is confirmation that the arc length formula is correct.

The **surface area** of an object is the area enclosed by the boundary between the object and its environment. For example, a cube is a shape whose length, width, and height are all equal and whose sides all meet each other at 90° angles. On its surface, a cube has six squares which all have the same size. The surface area of a cube is therefore equal to six times the area of one of the squares, or $A = 6L^2$.

The **cross-sectional area** of an object is the area that is created when the object is intersected by a single, flat plane. Imagine slicing through an object with a knife in one direction. One side of the interior of the object that is exposed along the cut would be the cross-sectional area. For example, a cylinder is a shape that has a constant circular cross-sectional area perpendicular to its height. The cross-sectional area of a cylinder perpendicular to its height is therefore the area of a circle, which is equal to π times the radius of the circle squared, or $A = \pi r^2$.

The **volume** of an object is the amount of three-dimensional space that it occupies. For example, the volume of a cube is the length of one of its sides cubed, or $V = L^3$, and the volume of a cylinder is its perpendicular cross-sectional area times its height, or $V = \pi r^2 h$.

0.5 | Trigonometry

Trigonometry is the geometry of triangles. You'll be using trigonometry frequently. Here is a review of the important relationships.

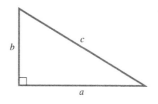

| **FIGURE 0.2**

The **Pythagorean theorem** is the relationship between the lengths of the three sides of a right triangle. (A right triangle is a triangle that has a 90° angle in it.) The longest side of the triangle is called the **hypotenuse.** The Pythagorean theorem says that the square of the hypotenuse is equal to the sum of the squares of the other two sides. (See Figure 0.2.) Symbolically,

$$c^2 = a^2 + b^2 \tag{0.3}$$

Trigonometric functions relate two sides of a triangle and one of the angles. There are six trigonometric functions: *sine, cosine, tangent, secant, cosecant,* and *cotangent.*

The **law of cosines** is a generalization of the Pythagorean theorem for any triangle, not just for right triangles. It looks similar to the Pythagorean theorem, with an extra term:

$$c^2 = a^2 + b^2 - 2ab \cos C \tag{0.4}$$

Angle C is opposite to side c. (See Figure 0.3.) Note that if angle C is 90°, then cos C is zero and the law of cosines reduces to the Pythagorean theorem.

The **Law of sines** gives ratios that are the same for any triangle:

$$\frac{(\sin A)}{a} = \frac{(\sin B)}{b} = \frac{(\sin C)}{c} \tag{0.5}$$

The angles A, B, and C are opposite to sides a, b, and c, respectively. Additionally, the three interior angles of any triangle always add up to 180°.

$$A + B + C = 180° \tag{0.6}$$

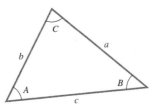

| **FIGURE 0.3**

0.6 | Equations, Symbols, and Units

Equations in mathematics are like sentences in English. They communicate fundamental mathematical ideas in a written form. When you write solutions to physics problems, you will almost always be writing in the language of mathematics. Be sure to use complete equations when you write.

A complete equation has three parts: (1) the left-hand side (LHS), (2) an equals sign ($=$), and (3) the right-hand side (RHS). Whenever you write an equation, you should make sure that the two things on either side of the equals sign are in fact the same.

In physics, the answer to a problem will usually be an equation. The following format should be used: Write the **symbol** of the quantity on the LHS and write what it is equal to on the RHS. In pure mathematics, what is on the RHS is just a number. In physics, however, the number must almost always be accompanied by one or more **units.** In most cases, a number without a unit is completely meaningless. Here is an example:

EXAMPLE
The Importance of Units

If you ask me how far my house is from campus and I answer "6," then have I really told you anything? No, because I have not specified the units. By specifying the units along with the number I will have nailed down the distance to a specific value. Here are some possibilities:

$$x = 6 \text{ miles}$$
$$x = 6 \text{ kilometers}$$
$$x = 6 \text{ blocks}$$

Therefore, in most cases, a number without units is completely meaningless. ∎

Most rules have exceptions, and this rule is no different. Some quantities have no units. However, a quantity without units is very rare. These rare cases will be noted as they arise.

Quantities are far superior to units for the following reason. Quantities such as distance, time, speed, force, and energy are determined by the physical structure of the universe in which we live. Units such as feet, miles, seconds, minutes, miles per hour, pounds, and ounces are completely arbitrary human constructs. Therefore, when discussing or writing about a particular event or condition, refer to the quantities, not the units. For example, ask, "What is the distance?" instead of "What are the miles?" *Units are arbitrary. Quantities are not.*

You'll be using **SI units.** SI stands for International System, or *Système International,* in French. Most people refer to this system as the metric system. A useful abbreviation for this system of units is MKS, which stands for meters, kilograms, and seconds. This abbreviation will be most useful because it will help us remember the units for length, mass, and time. Here is a list of the base units we will be using, along with their corresponding quantities in the MKS system:

Base MKS Unit	Quantity
meter	length
kilogram	mass
second	time
ampere	electric current
kelvin	temperature
mole	amount of substance

More complicated units can be formed by multiplying or dividing base units. For example, the MKS unit of force, the newton, is equal to kilograms times meters divided by seconds squared, or, in abbreviated form, $N = kg \cdot m/s^2$.

When writing very large or very small numbers, it is often useful to use a numeric prefix. These prefixes are letters that stand for large or small numbers and are used in order to write less. (As you will see, physicists and mathematicians like to write as little as possible.) *The most important thing to remember is that even though the prefixes are letters, they represent numbers and not units.* A list of numeric prefixes is provided in the Key Data section of this book.

Another way to write less is to use **scientific notation.** In this notation, a *nonzero* digit is written to the left of the decimal point, one or more digits are written to the right of the decimal point, and this quantity is multiplied by ten raised to some integer exponent.

EXAMPLE
Shorthand Notation

Suppose that the strength of a very small force is determined to be $F = 0.00512$ N. This value can be written using fewer symbols in two ways.

(a) One way would be to use scientific notation. In this form, the strength of the force would be $F = 5.12 \times 10^{-3}$ N.

(b) Another way would be to use a prefix. The prefix m stands for *milli,* which means "one one-thousandth" or 10^{-3}. Using a prefix, $F = 5.12$ mN.   ∎

0.7 | Unit Analysis

There are two different types of equations: those that relate quantities and those that relate units. Here is an example: Speed is equal to distance divided by elapsed time. The symbols for speed, distance, and elapsed time are, respectively, v, Δx, and Δt. The MKS units of speed, distance, and time are, respectively, meters per second, meters, and seconds.

$$v = \Delta x / \Delta t \text{ (quantity equation)}$$
$$(\text{m/s}) = (\text{m})/(\text{s}) \text{ (unit equation)}$$

In this case, the unit equation can be analyzed to check the quantity equation. But this is not always true. Here is another example: The area of a circle is equal to π times the radius of the circle squared. Imagine now that you make a mistake and forget the π. You write the area is equal to the radius squared. The quantity equation is wrong. However, the unit equation is still correct because π has no units. It is just a dimensionless number.

In the past you may have learned to check an equation by looking at the corresponding unit equation. This procedure is called **unit analysis.** This method is good, but there are better ways to go about checking your work. *Unit analysis can only tell you if a quantity equation is incorrect. It cannot tell you if it is correct.* Therefore, unit analysis is not stressed. Instead, we will focus on quantities.

0.8 | Significant Figures

The number of **significant figures** is related to the precision of a measurement or given value. The **first significant figure** is the one that is furthest to the left that is not zero.

Quantity and Numeric Value	First Significant Figure
$x = 51{,}234$ meters	5
$x = 0.051234$ meters	5
$x = 3{,}074{,}000$ meters	3

Notice that the location of the decimal point has nothing to do with significant figures.

EXAMPLE
Rounding

Answers are rounded to the appropriate number of significant figures. In the following example, the values are rounded to three significant figures.

Value Before Rounding	Value After Rounding
$t = 0.0000456889$ sec	$t = 0.0000457$ sec
$t = 123{,}400$ sec	$t = 123{,}000$ sec
$t = 0.101101$ sec	$t = 0.101$ sec

Notice again that the location of the decimal point has nothing to do with significant figures.   ∎

If the quantity is an integer, then the answer would never have a decimal point or any digits smaller than the ones digit. For example, if we determine that it takes ten people to lift a piano up off the ground, then we would write the answer as $N = 10$ people and not $N = 10.0$ people because it is absurd to refer to a non-integer number of people.

0.9 | Solving Simultaneous Equations

When solving physics problems, it is very rare to simply take a formula, put in numbers, and get another number. Typically, two or more relationships are considered. You will need to be able to combine the information in multiple relationships. Combining two or more relationships to get answers is called **solving simultaneous equations,** and it can be accomplished in three ways: *substitution, multiplication and subtraction*, or *forming a ratio*. Some techniques are easier than others, and some are more general. With practice, you will recognize which technique will be most useful in each situation. The three techniques are described in detail below.

Substitution is by far the most difficult and tedious method of the three. However, it is the most general; that is, it always works. The other two methods only work in certain cases. When in doubt, use the substitution method because it always works. Here are the steps:

1. Identify which equation is simplest.
2. Solve for one of the variables in the simplest equation.
3. Substitute the expression for that variable into one of the other equations.
4. Repeat if necessary until only one variable remains.
5. Solve the equation for the remaining variable using the rules of algebra.
6. Solve for the other variables.
7. Check the answers.

EXAMPLE
Using the Substitution Method

Suppose we have already determined that the following two relationships are important in determining the values of x and y. Now we want to combine the two relationships in order to determine the values of x and y that make both equations true.

$$3x + 2y = 7 \quad \text{and} \quad 5x - y = 3$$

1. The equation on the right is simpler because y is not being multiplied by a number.

2. Solve for y in the right equation:

$$y = 5x - 3$$

3. Substitute this expression into the other equation:

$$3x + 2(5x - 3) = 7$$

4. Since there is only one variable left (x) we do not need to repeat any steps.

5. Solve for x:

$$3x + 10x - 6 = 7$$
$$13x = 13$$
$$x = 1$$

6. Solve for y using $y = 5x - 3$:

$$y = 5(1) - 3 = 2$$

7. Check the answers. The right equation yields $7 = 7$ and the left one yields $3 = 3$ when the values of x and y are used. The answers $x = 1$ and $y = 2$ are correct.   ∎

0.10 | Scalar and Vector Quantities

Unless you have taken physics before, you have probably never heard of scalar or vector quantities. These terms refer to the two fundamental types of mathematical objects that we will be using. Whenever a new physical quantity is encountered, it will be important for you to know whether it is a scalar or a vector quantity.

A **scalar** quantity is simply a number. The number can be positive or negative. The scalar quantities we deal with will always be real numbers, never imaginary or complex numbers. If you find yourself taking the square root of a negative number, then you know that you have made a mistake because the result would be an imaginary or complex number.

A vector quantity is composed of two things: a number and a direction. The technical term for the number is **magnitude.** Other terms for the number are *size* and *amount*. The number can also be thought of as the *length* of the vector. However, not all vectors describe lengths so you must be very careful when thinking about a vector in terms of its length.

Many of the fundamental quantities in physics are vectors, not scalars. The exceptions are mass, time, energy, number of moles, electrical current, and quantities that are either derived from them or are the result of combining them. Power, for example, is a combination of energy and time. Therefore, power is also a scalar. Temperature is related to energy, so temperature is also a scalar.

Here is an example to help you understand the difference between vectors and scalars:

EXAMPLE
Speed and Velocity

Imagine driving from Los Angeles to San Francisco on the freeway. Speed is a scalar that would tell you how fast you are driving. The symbol for speed is v. If you were driving at 70 mph, then we would write $v = 70$ mph. On the other hand, velocity is a vector which would tell you how fast and also in which direction. The symbol for velocity is much the same as for speed, except that it has an arrow above it to indicate that it is a vector. In this example, your velocity would be written as $\vec{v} = (70$ mph, north$)$. If you were driving in the other direction, from San Francisco to Los Angeles, then your speed would be the same, but your velocity would be different. Your velocity now would be $\vec{v} = (70$ mph, south$)$.   ∎

We are concerned about direction because it will usually make a big difference. Imagine two people pushing equally hard on a heavy object that is initially not moving. If they push in the same direction, then the object might move if they push hard enough. But if they push in opposite directions, then, no matter how hard they push, the object will not move because the effects of their pushing will cancel out.

It is relatively easy to add and subtract scalars, and relatively difficult to multiply and divide them. For example, $113 + 23$ is easy; you can do it in your head, right? 113×23 is more difficult. You probably need a piece of paper and a pen or pencil to get the right answer.

Vectors are the exact opposite. It is relatively easy to multiply and divide vectors, and relatively difficult to add and subtract them. The techniques for adding and subtracting vectors will be outlined in the next two sections.

Vector Addition and Subtraction: Graphical Method

In practice, we will never add vectors graphically because the technique is not accurate enough for our needs, unless we use some fancy drafting software, which most of us don't have. However, it will often be helpful to get a rough idea using a pen or a pencil so that when we are finished adding or subtracting the vectors algebraically we can check our answer.

A vector will be drawn as an arrow; that is, as a straight line with an arrowhead at one end only. The end without the arrowhead is called the *tail*. The end with the arrowhead is called the *head*. This arrow points in the direction of the vector.

To add two vectors graphically, follow these steps. This is called the **tail-to-head method.** (See Figure 0.4.)

1. Draw a pair of coordinate axes (crossing horizontal and vertical lines). The point at which they cross is called the *origin*.
2. Draw the first vector ($\vec{A}$) with its tail at the origin.
3. Draw the second vector ($\vec{B}$) starting at the head of the first vector.
4. Draw a straight line connecting the origin to the head of the second vector.
5. The third line ($\vec{C}$) is the sum of the two vectors.

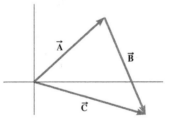

| FIGURE 0.4

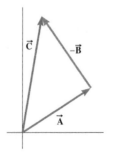

| FIGURE 0.5

To subtract one vector from another, multiply the vector to be subtracted ($\vec{\mathbf{B}}$) by -1. Now, add the first vector and the reversed second vector, using the steps above. The result will be the difference of the two vectors. (See Figure 0.5.)

In the next section we will learn how to add and subtract vectors using algebra. In practice, we will always add and subtract them this way.

Vector Addition and Subtraction: Component Method

There are two ways to describe a vector. The first way is called **polar form.** This is the way that the vectors will usually be given. It is also the way that the answer will usually be written. *Polar form gives the magnitude and the direction.* The direction is usually measured by starting on the horizontal line pointing to the right (the positive *x*-axis) and then rotating counterclockwise until you reach the vector. This is called measuring the angle in **standard polar form.**

The other way to describe a vector is called **Cartesian form,** named after René Descartes, the famous mathematician and philosopher. Cartesian form is used to add and subtract vectors, and it describes how much the vector points horizontally and vertically. These two aspects of a vector are called **components.** *Components that are to the right or up will be positive, while those to the left or down will be negative.* Here are two examples:

EXAMPLE
Converting from Polar to Cartesian Form

Imagine beginning a journey in New York City. You are going to fly along a straight line to Boston, which is 100 miles northeast from New York City. (Northeast is exactly halfway between north and east.) (See Figure 0.6.) In polar form, your

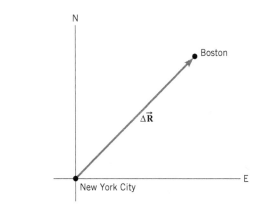

| FIGURE 0.6

displacement from New York City to Boston would be $\Delta\vec{R}$ (100 miles, 45.0°). What would your displacement be in Cartesian form? If we drew a diagram, made a right triangle, and used the definitions of sine and cosine, then we could determine the distance east and the distance north:

$$\Delta\vec{x} = \Delta R \cos\theta = 100 \cos 45.0° = +70.7 \text{ miles}$$

$$\Delta\vec{y} = \Delta R \sin\theta = 100 \sin 45.0° = +70.7 \text{ miles}$$

The displacement in Cartesian form would therefore be $\Delta\vec{R}$ = (70.7 miles, 70.7 miles). The components are equal because sine and cosine are equal when the angle is 45.0°.   ∎

EXAMPLE
Converting from Cartesian to Polar Form

The process also works in reverse. Imagine that a destination is 100 miles east and 50.0 miles north (see Figure 0.7). In Cartesian form, $\Delta\vec{R}$ = (100 miles, 50.0 miles). What would be the displacement in polar form? Again, if we drew the diagram and used trigonometry, then:

$$\Delta R = (\Delta x^2 + \Delta y^2)^{1/2} = (100^2 + 50.0^2)^{1/2} = 112 \text{ miles}$$

$$\theta = \tan^{-1}(\Delta y/\Delta x) = \tan^{-1}(50.0/100) = 26.6°$$

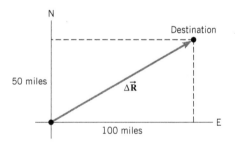

| FIGURE 0.7

The displacement in polar form would therefore be $\Delta\vec{R}$ = (112 miles, 26.6°).   ∎

Polar-to-Cartesian Transformation Equations

$$\Delta\vec{x} = \Delta R \cos\theta \qquad (0.7)$$

$$\Delta\vec{y} = \Delta R \sin\theta \qquad (0.8)$$

Cartesian-to-Polar Transformation Equations

$$\Delta R = (\Delta x^2 + \Delta y^2)^{1/2} \tag{0.9}$$

$$\theta \;\; = \tan^{-1}(\Delta y/\Delta x) \tag{0.10}$$

Note: In order for equations 0.7, 0.8, and 0.10 to work correctly, the angle θ must be expressed in standard polar form.

Now that we know how to change the form of a vector, we are ready to learn how to add and subtract them algebraically.

To add two vectors algebraically, follow these steps. This is called the **component method.**

1. Use sine and cosine to compute the horizontal and vertical components of the vectors.

2. Add the two horizontal components together. Do the same for the vertical ones.

3. Convert back to polar form for the final answer.

To subtract two vectors algebraically, do the same steps above, except in step two subtract the components instead of adding them.

EXAMPLE
Vector Addition Using the Component Method

If I walk northeast for 10.0 miles and then southeast for 5.00 miles, then how far am I from the starting point? (See Figure 0.8.) The answer is not 15.0 miles because I did not walk in the same direction the whole time. Following the steps above, I can determine the total distance.

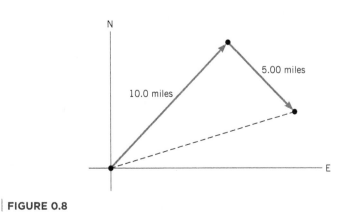

| FIGURE 0.8

$$\Delta \vec{R}_1 = (10.0 \text{ miles}, 45.0°) \text{ and } \Delta \vec{R}_2 = (5.00 \text{ miles}, 315°)$$

where the angles are given in standard polar form.

$$\Delta \vec{x}_1 = \Delta R_1 \cos \theta_1 \qquad = 10.0 \cos 45.0° \qquad = +7.07106 \text{ miles}$$

$$\Delta \vec{x}_2 = \Delta R_2 \cos \theta_2 \qquad = 5.00 \cos 315° \qquad = +3.53553 \text{ miles}$$

$$\Delta \vec{x} = 7.07106 + 3.53553 = 10.60659 \text{ miles}$$

$$\Delta \vec{y}_1 = \Delta R_1 \sin \theta_1 \qquad = 10.0 \sin 45.0° \qquad = +7.07106 \text{ miles}$$

$$\Delta \vec{y}_2 = \Delta R_2 \sin \theta_2 \qquad = 5.00 \sin 315° \qquad = -3.53553 \text{ miles}$$

$$\Delta \vec{y} = 7.07106 + (-3.53553) = 3.53553 \text{ miles}$$

$$\Delta R = (\Delta x^2 + \Delta y^2)^{1/2} = (10.60659^2 + 3.53553^2)^{1/2} = 11.2 \text{ miles}$$

Notice that the horizontal components combine to make a larger total. This is because I walked east all the time, never west. Also notice that the vertical components partially cancel. This is because some of the time I walked north and some of the time I walked south. ∎

0.11 | Exponents and Logarithms

An **exponential equation** is a compact way of writing a number multiplied by itself any number of times. The following notation is used:

$$\textbf{base}^{\textbf{exponent}} = \textbf{answer}$$

The base is the number being multiplied by itself, the exponent is the number of times the base is multiplied, and the answer is the result of the multiplication. For example, $5^3 = 5 \times 5 \times 5 = 125$. If the exponent is one, then the answer is equal to the base. For example $4^1 = 4$. If the exponent is zero, then the answer is one. For example, $6^0 = 1$. If the exponent is negative, then the answer is one divided by what the answer would be if the exponent was positive. For example,
$$5^{-2} = \frac{1}{5^2} = \frac{1}{25} = 0.04.$$
The following properties of exponents can come in handy from time to time:

$$A^x A^y = A^{x+y} \text{ (Add the exponents together.)} \tag{0.11}$$

$$\frac{A^x}{A^y} = A^{x-y} \text{ (Subtract one exponent from the other.)} \tag{0.12}$$

$$(A^x)^y = A^{xy} \quad \text{(Multiply the exponents together.)} \tag{0.13}$$

A logarithmic equation is another way of expressing the same information contained in an exponential equation. Technically, exponential and logarithmic functions are inverse functions of each other. The following notation is used for logarithmic equations:

$$\log_{\text{base}} \text{answer} = \text{exponent}$$

It is often useful to rewrite logarithmic equations as exponential equations, and vice versa. For example, $\log_{10} 100 = 2$ is equivalent to $10^2 = 100$. If the answer is equal to the base, then the exponent is one. For example $\log_4 4 = 1$. If the answer is one, then the exponent is zero. For example, $\log_6 1 = 0$.

The following properties of logarithms can come in handy from time to time:

$$\log_x(AB) = (\log_x A) + (\log_x B) \quad \text{(Add the logarithms together.)} \tag{0.14}$$

$$\log_x\left(\frac{A}{B}\right) = (\log_x A) - (\log_x B) \quad \text{(Subtract one logarithm from the other.)} \tag{0.15}$$

$$\log_x(A^N) = N(\log_x A) \qquad \text{(Multiply the logarithm and the exponent.)} \tag{0.16}$$

$$\log_x A = \frac{\log_b A}{\log_b x} \qquad \text{(Change from one base to another.)} \tag{0.17}$$

CHAPTER 1

Introduction and Mathematical Concepts

1.2 | Units

The SI system of units includes the meter (m), the kilogram (kg), and the second (s) as the base units for length, mass, and time, respectively.

One meter is the distance that light travels in a vacuum in a time of 1/299 792 458 second.

One kilogram is the mass of a standard cylinder of platinum–iridium alloy kept at the International Bureau of Weights and Measures.

One second is the time for a certain type of electromagnetic wave emitted by cesium-133 atoms to undergo 9 192 631 770 wave cycles.

1.3 | The Role of Units in Problem Solving

To convert a number from one unit to another, multiply the number by the ratio of the two units. For instance, to convert 979 meters to feet, multiply 979 meters by the factor (3.281 foot/1 meter).

The dimension of a quantity represents its physical nature and the type of unit used to specify it. Three such dimensions are length [L], mass [M], time [T].

Dimensional analysis is a method for checking mathematical relations for the consistency of their dimensions.

1.4 | Trigonometry

The sine, cosine, and tangent functions of an angle θ are defined in terms of a right triangle that contains θ:

$$\sin \theta = \frac{h_o}{h} \quad \textbf{(1.1)} \qquad \cos \theta = \frac{h_a}{h} \quad \textbf{(1.2)} \qquad \tan \theta = \frac{h_o}{h_a} \quad \textbf{(1.3)}$$

where h_o and h_a are, respectively, the lengths of the sides opposite and adjacent to the angle θ, and h is the length of the hypotenuse.

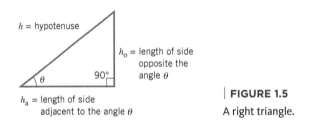

h = hypotenuse

h_o = length of side opposite the angle θ

90°

h_a = length of side adjacent to the angle θ

| **FIGURE 1.5**
A right triangle.

The inverse sine, inverse cosine, and inverse tangent functions are

$$\theta = \sin^{-1}\left(\frac{h_o}{h}\right) \quad \textbf{(1.4)} \qquad \theta = \cos^{-1}\left(\frac{h_a}{h}\right) \quad \textbf{(1.5)} \qquad \theta = \tan^{-1}\left(\frac{h_o}{h_a}\right) \quad \textbf{(1.6)}$$

The Pythagorean theorem states that the square of the length of the hypotenuse of a right triangle is equal to the sum of the squares of the lengths of the other two sides:

$$h^2 = h_{\mathrm{o}}^2 + h_{\mathrm{a}}^2 \tag{1.7}$$

1.5 | Scalars and Vectors

A scalar quantity is described completely by its size, which is also called its magnitude. A vector quantity has both a magnitude and a direction. Vectors are often represented by arrows, the length of the arrow being proportional to the magnitude of the vector and the direction of the arrow indicating the direction of the vector.

1.6 | Vector Addition and Subtraction

One procedure for adding vectors utilizes a graphical technique, in which the vectors to be added are arranged in a tail-to-head fashion. The resultant vector is drawn from the tail of the first vector to the head of the last vector.

The subtraction of a vector is treated as the addition of a vector that has been multiplied by a scalar factor of -1. Multiplying a vector by -1 reverses the direction of the vector.

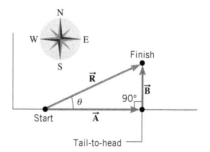

| **FIGURE 1.10**

The addition of two perpendicular displacement vectors $\vec{\mathbf{A}}$ and $\vec{\mathbf{B}}$ gives the resultant vector $\vec{\mathbf{R}}$.

1.7 | The Components of a Vector

In two dimensions, the vector components of a vector $\vec{\mathbf{A}}$ are two perpendicular vectors $\vec{\mathbf{A}}_x$ and $\vec{\mathbf{A}}_y$ that are parallel to the x and y axes, respectively, and that add together vectorially so that $\vec{\mathbf{A}} = \vec{\mathbf{A}}_x + \vec{\mathbf{A}}_y$.

The scalar component A_x has a magnitude that is equal to that of $\vec{\mathbf{A}}_x$ and is given a positive sign if $\vec{\mathbf{A}}_x$ points along the $+x$ axis and a negative sign if $\vec{\mathbf{A}}_x$ points along the $-x$ axis. The scalar component A_y is defined in a similar manner.

A vector is zero if, and only if, each of its vector components is zero.

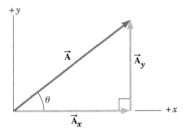

| FIGURE 1.16
An arbitrary vector $\vec{A}$ and its vector components $\vec{A}_x$ and $\vec{A}_y$.

Two vectors are equal if, and only if, they have the same magnitude and direction. Alternatively, two vectors are equal in two dimensions if the x vector components of each are equal and the y vector components of each are equal.

1.8 | Addition of Vectors by Means of Components

If two vectors $\vec{A}$ and $\vec{B}$ are added to give a resultant $\vec{C}$ such that $\vec{C} = \vec{A} + \vec{B}$, then

$$C_x = A_x + B_x \qquad \text{and} \qquad C_y = A_y + B_y$$

where C_x, A_x, and B_x are the scalar components of the vectors along the x direction, and C_y, A_y, and B_y are the scalar components of the vectors along the y direction.

REASONING STRATEGY
The Component Method of Vector Addition

1. For each vector to be added, determine the x and y components relative to a conveniently chosen x, y coordinate system. Be sure to take into account the directions of the components by using plus and minus signs to denote whether the components point along the positive or negative axes.

2. Find the algebraic sum of the x components, which is the x component of the resultant vector. Similarly, find the algebraic sum of the y components, which is the y component of the resultant vector.

3. Use the x and y components of the resultant vector and the Pythagorean theorem to determine the magnitude of the resultant vector.

4. Use either the inverse sine, inverse cosine, or inverse tangent function to find the angle that specifies the direction of the resultant vector.

CHAPTER 2

Kinematics in One Dimension

2.1 | Displacement

Displacement is a vector that points from an object's initial position to its final position. The magnitude of the displacement is the shortest distance between the two positions.

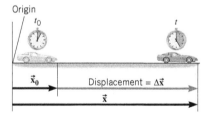

| FIGURE 2.1
The displacement $\Delta\vec{x}$ is a vector that points from the initial position $\vec{x}_0$ to the final position $\vec{x}$.

2.2 | Speed and Velocity

The average speed of an object is the distance traveled by the object divided by the time required to cover the distance:

$$\text{Average speed} = \frac{\text{Distance}}{\text{Elapsed time}} \tag{2.1}$$

The average velocity $\vec{\overline{v}}$ of an object is the object's displacement $\Delta \vec{x}$ divided by the elapsed time Δt:

$$\vec{\overline{v}} = \frac{\Delta \vec{x}}{\Delta t} \qquad (2.2)$$

Average velocity is a vector that has the same direction as the displacement. When the elapsed time becomes infinitesimally small, the average velocity becomes equal to the instantaneous velocity $\vec{v}$, the velocity at an instant of time:

$$\vec{v} = \lim_{\Delta t \to 0} \frac{\Delta \vec{x}}{\Delta t} \qquad (2.3)$$

2.3 | Acceleration

The average acceleration $\vec{\overline{a}}$ is a vector. It equals the change $\Delta \vec{v}$ in the velocity divided by the elapsed time Δt, the change in the velocity being the final minus the initial velocity:

$$\vec{\overline{a}} = \frac{\Delta \vec{v}}{\Delta t} \qquad (2.4)$$

When Δt becomes infinitesimally small, the average acceleration becomes equal to the instantaneous acceleration $\vec{a}$:

$$\vec{a} = \lim_{\Delta t \to 0} \frac{\Delta \vec{v}}{\Delta t} \qquad (2.5)$$

Acceleration is the rate at which the velocity is changing.

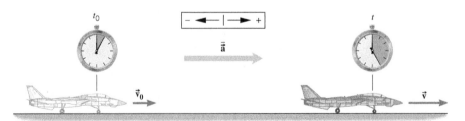

| FIGURE 2.4
During takeoff, the plane accelerates from an initial velocity $\vec{v}_0$ to a final velocity $\vec{v}$ during the time interval $\Delta t = t - t_0$.

2.4 | Equations of Kinematics for Constant Acceleration;
2.5 | Applications of the Equations of Kinematics

The equations of kinematics apply when an object moves with a constant acceleration along a straight line. These equations relate the displacement $x - x_0$, the acceleration a, the final velocity v, the initial velocity v_0, and the elapsed time $t - t_0$. Assuming that $x_0 = 0$ m at $t_0 = 0$ s, the equations of kinematics are

$$v = v_0 + at \qquad (2.4)$$
$$x = \tfrac{1}{2}(v_0 + v)t \qquad (2.7)$$
$$x = v_0 t + \tfrac{1}{2}at^2 \qquad (2.8)$$
$$v^2 = v_0^2 + 2ax \qquad (2.9)$$

2.6 | Freely Falling Bodies

In free-fall motion, an object experiences negligible air resistance and a constant acceleration due to gravity. All objects at the same location above the earth have the same acceleration due to gravity. The acceleration due to gravity is directed toward the center of the earth and has a magnitude of approximately 9.80 m/s² near the earth's surface.

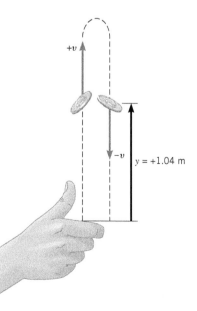

$y = +1.04$ m

| FIGURE 2.16

For a given displacement along the motional path, the upward speed of the coin is equal to its downward speed, but the two velocities point in opposite directions.

2.7 | Graphical Analysis of Velocity and Acceleration

The slope of a plot of position versus time for a moving object gives the object's velocity. The slope of a plot of velocity versus time gives the object's acceleration.

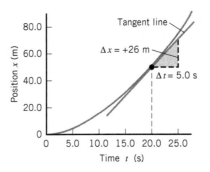

| FIGURE 2.20
When the velocity is changing, the position-vs.-time graph is a curved line. The slope $\Delta x / \Delta t$ of the tangent line drawn to the curve at a given time is the instantaneous velocity at that time.

CHAPTER 3

Kinematics in Two Dimensions

3.1 | Displacement, Velocity, and Acceleration

The position of an object is located with a vector $\vec{r}$ drawn from the coordinate origin to the object. The displacement $\Delta\vec{r}$ of the object is defined as $\Delta\vec{r} = \vec{r} - \vec{r}_0$, where $\vec{r}$ and $\vec{r}_0$ specify its final and initial positions, respectively.

The average velocity $\overline{\vec{v}}$ of an object moving between two positions is defined as its displacement $\Delta\vec{r} = \vec{r} - \vec{r}_0$ divided by the elapsed time $\Delta t = t - t_0$:

$$\overline{\vec{v}} = \frac{\vec{r} - \vec{r}_0}{t - t_0} = \frac{\Delta\vec{r}}{\Delta t} \tag{3.1}$$

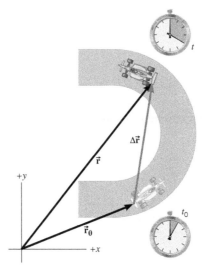

| FIGURE 3.1

The displacement $\Delta\vec{r}$ of the car is a vector that points from the initial position of the car at time t_0 to the final position at time t. The magnitude of $\Delta\vec{r}$ is the shortest distance between the two positions.

The instantaneous velocity $\vec{\mathbf{v}}$ is the velocity at an instant of time. The average velocity becomes equal to the instantaneous velocity in the limit that Δt becomes infinitesimally small $(\Delta t \rightarrow 0 \text{ s})$:

$$\vec{\mathbf{v}} = \lim_{\Delta t \to 0} \frac{\Delta \vec{\mathbf{r}}}{\Delta t}$$

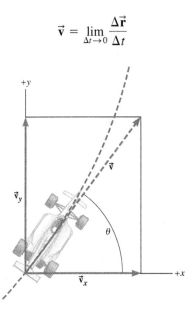

| **FIGURE 3.2**
The instantaneous velocity $\vec{\mathbf{v}}$ and its two vector components $\vec{\mathbf{v}}_x$ and $\vec{\mathbf{v}}_y$.

The average acceleration $\vec{\mathbf{a}}$ of an object is the change in its velocity, $\Delta \vec{\mathbf{v}} = \vec{\mathbf{v}} - \vec{\mathbf{v}}_0$ divided by the elapsed time $\Delta t = t - t_0$:

$$\vec{\mathbf{a}} = \frac{\vec{\mathbf{v}} - \vec{\mathbf{v}}_0}{t - t_0} = \frac{\Delta \vec{\mathbf{v}}}{\Delta t} \tag{3.2}$$

The instantaneous acceleration $\vec{\mathbf{a}}$ is the acceleration at an instant of time. The average acceleration becomes equal to the instantaneous acceleration in the limit that the elapsed time Δt becomes infinitesimally small:

$$\vec{\mathbf{a}} = \lim_{\Delta t \to 0} \frac{\Delta \vec{\mathbf{v}}}{\Delta t}$$

3.2 | Equations of Kinematics in Two Dimensions

Motion in two dimensions can be described in terms of the time t and the x and y components of four vectors: the displacement, the acceleration, and the initial and final velocities.

The x part of the motion occurs exactly as it would if the y part did not occur at all. Similarly, the y part of the motion occurs exactly as it would if the x part of the motion did not exist. The motion can be analyzed by treating the x and y components of the four vectors separately and realizing that the time t is the same for each component.

When the acceleration is constant, the x components of the displacement, the acceleration, and the initial and final velocities are related by the equations of kinematics, and so are the y components:

x Component		y Component	
$v_x = v_{0x} + a_x t$	(3.3a)	$v_y = v_{0y} + a_y t$	(3.3b)
$x = \frac{1}{2}(v_{0x} + v_x)t$	(3.4a)	$y = \frac{1}{2}(v_{0y} + v_y)t$	(3.4b)
$x = v_{0x}t + \frac{1}{2}a_x t^2$	(3.5a)	$y = v_{0y}t + \frac{1}{2}a_y t^2$	(3.5b)
$v_x^2 = v_{0x}^2 + 2a_x x$	(3.6a)	$v_y^2 = v_{0y}^2 + 2a_y y$	(3.6b)

The directions of these components are conveyed by assigning a plus ($+$) or minus ($-$) sign to each one.

3.3 | Projectile Motion

Projectile motion is an idealized kind of motion that occurs when a moving object (the projectile) experiences only the acceleration due to gravity, which acts vertically downward. If the trajectory of the projectile is near the earth's surface, a_y has a magnitude of 9.80 m/s^2. The acceleration has no horizontal component ($a_x = 0$ m/s^2), the effects of air resistance being negligible.

There are several symmetries in projectile motion: (1) The time to reach maximum height from any point is equal to the time spent returning from the maximum height to that point. (2) The speed of a projectile depends only on its height above its launch point, and not on whether it is moving upward or downward.

3.4 | Relative Velocity

The velocity of object A relative to object B is $\vec{v}_{AB}$, and the velocity of object B relative to object C is $\vec{v}_{BC}$. The velocity of A relative to C is (note the ordering of the subscripts)

$$\vec{v}_{AC} = \vec{v}_{AB} + \vec{v}_{BC}$$

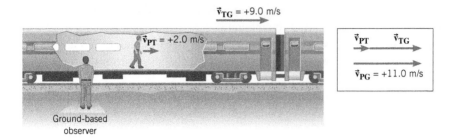

| FIGURE 3.15

The velocity of the passenger relative to the ground-based observer is $\vec{v}_{PG}$. It is the vector sum of the velocity $\vec{v}_{PT}$ of the passenger relative to the train and the velocity $\vec{v}_{TG}$ of the train relative to the ground: $\vec{v}_{PG} = \vec{v}_{PT} + \vec{v}_{TG}$.

While the velocity of object A relative to object B is $\vec{v}_{AB}$, the velocity of B relative to A is $\vec{v}_{BA} = -\vec{v}_{AB}$.

CHAPTER 4

Forces and Newton's Laws of Motion

4.1 | The Concepts of Force and Mass

A force is a push or a pull and is a vector quantity. Contact forces arise from the physical contact between two objects. Noncontact forces are also called action-at-a-distance forces, because they arise without physical contact between two objects.

Mass is a property of matter that determines how difficult it is to accelerate or decelerate an object. Mass is a scalar quantity.

4.2 | Newton's First Law of Motion

Newton's first law of motion, sometimes called the law of inertia, states that an object continues in a state of rest or in a state of motion at a constant velocity unless compelled to change that state by a net force.

Inertia is the natural tendency of an object to remain at rest or in motion at a constant velocity. The mass of a body is a quantitative measure of inertia and is measured in an SI unit called the kilogram (kg). An inertial reference frame is one in which Newton's law of inertia is valid.

4.3 | Newton's Second Law of Motion;
4.4 | The Vector Nature
of Newton's Second Law of Motion

Newton's second law of motion states that when a net force $\Sigma\vec{F}$ acts on an object of mass m, the acceleration $\vec{a}$ of the object can be obtained from the following equation:

$$\Sigma\vec{F} = m\vec{a} \tag{4.1}$$

This is a vector equation and, for motion in two dimensions, is equivalent to the following two equations:

$$\Sigma F_x = ma_x \tag{4.2a}$$

$$\Sigma F_y = ma_y \tag{4.2b}$$

In these equations the x and y subscripts refer to the scalar components of the force and acceleration vectors. The SI unit of force is the Newton (N).

When determining the net force, a free-body diagram is helpful. A free-body diagram is a diagram that represents the object and the forces acting on it.

4.5 | Newton's Third Law of Motion

Newton's third law of motion, often called the action–reaction law, states that whenever one object exerts a force on a second object, the second object exerts an oppositely directed force of equal magnitude on the first object.

4.6 | Types of Forces: An Overview

Only three fundamental forces have been discovered: the gravitational force, the strong nuclear force, and the electroweak force. The electroweak force manifests itself as either the electromagnetic force or the weak nuclear force.

4.7 | The Gravitational Force

Newton's law of universal gravitation states that every particle in the universe exerts an attractive force on every other particle. For two particles that are separated by a distance r and have masses m_1 and m_2, the law states that the magnitude of this attractive force is

$$F = G\frac{m_1 m_2}{r^2}$$
(4.3)

The direction of this force lies along the line between the particles. The constant G has a value of $G = 6.674 \times 10^{-11}$ N $\cdot$ m^2/kg^2 and is called the universal gravitational constant.

The weight W of an object on or above the earth is the gravitational force that the earth exerts on the object and can be calculated from the mass m of the object and the magnitude g of the acceleration due to the earth's gravity according to

$$W = mg$$
(4.5)

4.8 | The Normal Force

The normal force $\vec{F}_N$ is one component of the force that a surface exerts on an object with which it is in contact—namely, the component that is perpendicular to the surface.

The apparent weight is the force that an object exerts on the platform of a scale and may be larger or smaller than the true weight mg if the object and the scale have an acceleration a (+ if upward, − if downward). The apparent weight is

$$\text{Apparent weight} = mg + ma$$
(4.6)

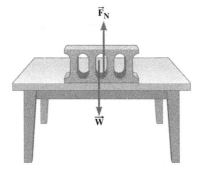

| FIGURE 4.13
Two forces act on the block, its weight $\vec{W}$ and the normal force $\vec{F}_N$ exerted by the surface of the table.

4.9 | Static and Kinetic Frictional Forces

A surface exerts a force on an object with which it is in contact. The component of the force perpendicular to the surface is called the normal force. The component parallel to the surface is called friction.

The force of static friction between two surfaces opposes any impending relative motion of the surfaces. The magnitude of the static frictional force depends on the magnitude of the applied force and can assume any value up to a maximum of

$$f_s^{\text{MAX}} = \mu_s F_N \tag{4.7}$$

where μ_s is the coefficient of static friction and F_N is the magnitude of the normal force.

The force of kinetic friction between two surfaces sliding against one another opposes the relative motion of the surfaces. This force has a magnitude given by

$$f_k = \mu_k F_N \tag{4.8}$$

where μ_k is the coefficient of kinetic friction.

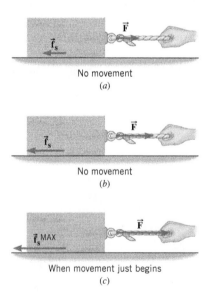

No movement
(a)

No movement
(b)

When movement just begins
(c)

| FIGURE 4.20
Applying a small force $\vec{F}$ to the block, as in parts *a* and *b*, produces no movement, because the static frictional force $\vec{f}_s$ exactly balances the applied force. (*c*) The block just begins to move when the applied force is slightly greater than the maximum static frictional force $\vec{f}_s^{\text{MAX}}$.

TABLE 4.2

Approximate Values of the Coefficients of Friction for Various Surfaces

Materials	Coefficient of Static Friction, μ_s	Coefficient of Kinetic Friction, μ_k
Glass on glass (dry)	0.94	0.4
Ice on ice (clean, 0°C)	0.1	0.02
Rubber on dry concrete	1.0	0.8
Rubber on wet concrete	0.7	0.5
Steel on ice	0.1	0.05
Steel on steel (dry hard steel)	0.78	0.42
Teflon on Teflon	0.04	0.04
Wood on wood	0.35	0.3

4.10 | The Tension Force

The word "tension" is commonly used to mean the tendency of a rope to be pulled apart due to forces that are applied at each end. Because of tension, a rope transmits a force from one end to the other. When a rope is accelerating, the force is transmitted undiminished only if the rope is massless.

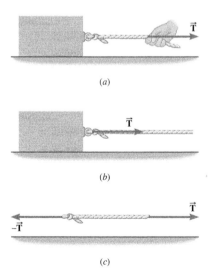

(a)

(b)

(c)

| **FIGURE 4.25**
(a) A force $\vec{T}$ is being applied to the right end of a rope.
(b) The force is transmitted to the box.
(c) Forces are applied to both ends of the rope. These forces have equal magnitudes and opposite directions.

4.11 | Equilibrium Applications of Newton's Laws of Motion

An object is in equilibrium when the object has zero acceleration, or, in other words, when it moves at a constant velocity (which may be zero). The sum of the forces that act on an object in equilibrium is zero. Under equilibrium conditions in two dimensions, the separate sums of the force components in the x direction and in the y direction must each be zero:

$$\Sigma F_x = 0 \tag{4.9a}$$

$$\Sigma F_y = 0 \tag{4.9b}$$

4.12 | Nonequilibrium Applications of Newton's Laws of Motion

If an object is not in equilibrium, then Newton's second law must be used to account for the acceleration:

$$\Sigma F_x = ma_x \tag{4.2a}$$

$$\Sigma F_y = ma_y \tag{4.2b}$$

CHAPTER 5

Dynamics of Uniform Circular Motion

5.1 | Uniform Circular Motion

Uniform circular motion is the motion of an object traveling at a constant (uniform) speed on a circular path.

The period T is the time required for the object to travel once around the circle. The speed v of the object is related to the period and the radius r of the circle by

$$v = \frac{2\pi r}{T} \tag{5.1}$$

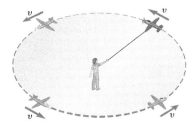

| FIGURE 5.1

The motion of a model airplane flying at a constant speed on a horizontal circular path is an example of uniform circular motion.

5.2 │ Centripetal Acceleration

An object in uniform circular motion experiences an acceleration, known as centripetal acceleration. The magnitude a_c of the centripetal acceleration is

$$a_c = \frac{v^2}{r} \qquad\qquad (5.2)$$

where v is the speed of the object and r is the radius of the circle. The direction of the centripetal acceleration vector always points toward the center of the circle and continually changes as the object moves.

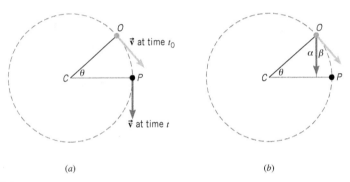

(a)   (b)

│ **FIGURE 5.2**
(a) For an object (•) in uniform circular motion, the velocity $\vec{v}$ has different directions at different places on the circle.
(b) The velocity vector has been removed from point P, shifted parallel to itself, and redrawn with its tail at point O.

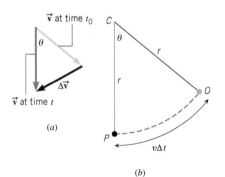

(b)

│ **FIGURE 5.3**
(a) The directions of the velocity vector at times t and t_0 differ by the angle θ.
(b) When the object moves along the circle from O to P, the radius r traces out the same angle θ. Here, the sector COP has been rotated clockwise by 90° relative to its orientation in Figure 5.2.

5.3 | Centripetal Force

To produce a centripetal acceleration, a net force pointing toward the center of the circle is required. This net force is called the centripetal force, and its magnitude F_c is

$$F_c = \frac{mv^2}{r} \tag{5.3}$$

where m and v are the mass and speed of the object, and r is the radius of the circle. The direction of the centripetal force vector, like that of the centripetal acceleration vector, always points toward the center of the circle.

5.4 | Banked Curves

A vehicle can negotiate a circular turn without relying on static friction to provide the centripetal force, provided the turn is banked at an angle relative to the horizontal. The angle θ at which a friction-free curve must be banked is related to the speed v of

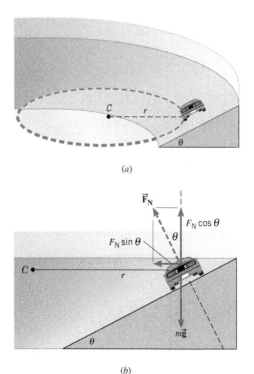

(a)

(b)

| FIGURE 5.11

(a) A car travels on a circle of radius r on a frictionless banked road. The banking angle is θ, and the center of the circle is at C.

(b) The forces acting on the car are its weight $m\vec{g}$ and the normal force $\vec{F}_N$. A component $F_N \sin \theta$ of the normal force provides the centripetal force.

the vehicle, the radius r of the curve, and the magnitude g of the acceleration due to gravity by

$$\tan \theta = \frac{v^2}{rg} \qquad (5.4)$$

5.5 | Satellites in Circular Orbits

When a satellite orbits the earth, the gravitational force provides the centripetal force that keeps the satellite moving in a circular orbit. The speed v and period T of a satellite depend on the mass M_E of the earth and the radius r of the orbit according to

$$v = \sqrt{\frac{GM_E}{r}} \qquad (5.5)$$

$$T = \frac{2\pi r^{3/2}}{\sqrt{GM_E}} \qquad (5.6)$$

where G is the universal gravitational constant.

5.6 | Apparent Weightlessness and Artificial Gravity

The apparent weight of an object is the force that it exerts on a scale with which it is in contact. All objects, including people, on board an orbiting satellite are in free-fall, since they experience negligible air resistance and they have an acceleration that is equal to the acceleration due to gravity. When a person is in free-fall, his or her apparent weight is zero, because both the person and the scale fall freely and cannot push against one another.

5.7 | Vertical Circular Motion

Vertical circular motion occurs when an object, such as a motorcycle, moves on a vertical circular path. The speed of the object often varies from moment to moment, and so do the magnitudes of the centripetal acceleration and centripetal force.

CHAPTER 6

Work and Energy

6.1 | Work Done by a Constant Force

The work W done by a constant force acting on an object is

$$W = (F \cos \theta)s \tag{6.1}$$

where F is the magnitude of the force, s is the magnitude of the displacement, and θ is the angle between the force and the displacement vectors. Work is a scalar quantity and can be positive or negative, depending on whether the force has a component that points, respectively, in the same direction as the displacement or in the opposite direction. The work is zero if the force is perpendicular ($\theta = 90°$) to the displacement.

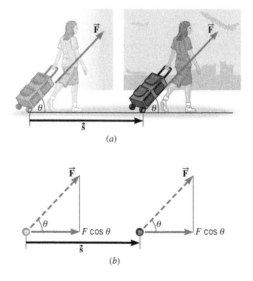

| FIGURE 6.2

(a) Work can be done by a force $\vec{F}$ that points at an angle θ relative to the displacement $\vec{s}$.
(b) The force component that points along the displacement is $F \cos \theta$.

6.2 | The Work–Energy Theorem and Kinetic Energy

The kinetic energy KE of an object of mass m and speed v is

$$\text{KE} = \frac{1}{2}mv^2 \qquad (6.2)$$

The work–energy theorem states that the work W done by the net external force acting on an object equals the difference between the object's final kinetic energy KE_f and initial kinetic energy KE_0:

$$W = \text{KE}_f - \text{KE}_0 \qquad (6.3)$$

The kinetic energy increases when the net force does positive work and decreases when the net force does negative work.

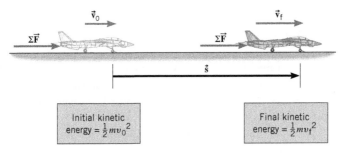

| FIGURE 6.5
A constant net external force $\Sigma\vec{F}$ acts over a displacement $\vec{s}$ and does work on the plane. As a result of the work done, the plane's kinetic energy changes.

6.3 | Gravitational Potential Energy

The work done by the force of gravity on an object of mass m is

$$W_{\text{gravity}} = mg(h_0 - h_f) \qquad (6.4)$$

where h_0 and h_f are the initial and final heights of the object, respectively.

Gravitational potential energy PE is the energy that an object has by virtue of its position. For an object near the surface of the earth, the gravitational potential energy is given by

$$\text{PE} = mgh \qquad (6.5)$$

where h is the height of the object relative to an arbitrary zero level.

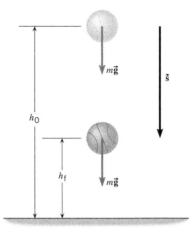

| FIGURE 6.9
Gravity exerts a force $m\vec{g}$ on the basketball. Work is done by the gravitational force as the basketball falls from a height of h_0 to a height of h_f.

6.4 | Conservative Versus Nonconservative Forces

A conservative force is one that does the same work in moving an object between two points, independent of the path taken between the points. Alternatively, a force is conservative if the work it does in moving an object around any closed path is zero. A force is nonconservative if the work it does on an object moving between two points depends on the path of the motion between the points.

6.5 | The Conservation of Mechanical Energy

The total mechanical energy E is the sum of the kinetic energy and potential energy:

$$E = \mathrm{KE} + \mathrm{PE}$$

The work–energy theorem can be expressed in an alternate form as

$$W_{\mathrm{nc}} = E_{\mathrm{f}} - E_0 \tag{6.8}$$

where W_{nc} is the net work done by the external nonconservative forces, and E_{f} and E_0 are the final and initial total mechanical energies, respectively.

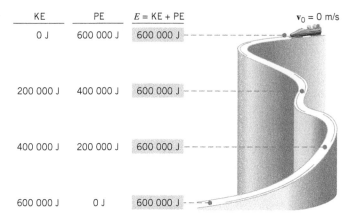

KE	PE	E = KE + PE
0 J	600 000 J	600 000 J
200 000 J	400 000 J	600 000 J
400 000 J	200 000 J	600 000 J
600 000 J	0 J	600 000 J

$v_0 = 0$ m/s

| FIGURE 6.15

If friction and wind resistance are ignored, a bobsled run illustrates how kinetic and potential energy can be interconverted, while the total mechanical energy remains constant. The total mechanical energy is 600 000 J, being all potential energy at the top and all kinetic energy at the bottom.

The principle of conservation of mechanical energy states that the total mechanical energy E remains constant along the path of an object, provided that the net work done by external nonconservative forces is zero. Whereas E is constant, KE and PE may be transformed into one another.

6.6 | Nonconservative Forces and the Work–Energy Theorem; 6.7 | Power

Average power $\overline{P}$ is the work done per unit time or the rate at which work is done:

$$\overline{P} = \frac{\text{Work}}{\text{Time}} \qquad (6.10a)$$

It is also the rate at which energy changes:

$$\overline{P} = \frac{\text{Change in energy}}{\text{Time}} \qquad (6.10b)$$

When a force of magnitude F acts on an object moving with an average speed $\overline{v}$, the average power is given by

$$\overline{P} = F\,\overline{v} \qquad (6.11)$$

6.8 | Other Forms of Energy and the Conservation of Energy

The principle of conservation of energy states that energy cannot be created or destroyed but can only be transformed from one form to another.

6.9 | Work Done by a Variable Force

The work done by a variable force of magnitude F in moving an object through a displacement of magnitude s is equal to the area under the graph of $F \cos \theta$ versus s. The angle θ is the angle between the force and displacement vectors.

CHAPTER 7

Impulse and Momentum

7.1 | The Impulse–Momentum Theorem

The impulse $\vec{\mathbf{J}}$ of a force is the product of the average force $\vec{\overline{\mathbf{F}}}$ and the time interval Δt during which the force acts:

$$\vec{\mathbf{J}} = \vec{\overline{\mathbf{F}}}\Delta t \tag{7.1}$$

Impulse is a vector that points in the same direction as the average force.

The linear momentum $\vec{\mathbf{p}}$ of an object is the product of the object's mass m and velocity $\vec{\mathbf{v}}$:

$$\vec{\mathbf{p}} = m\vec{\mathbf{v}} \tag{7.2}$$

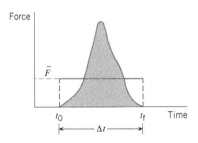

| FIGURE 7.1

When a bat strikes a ball, the magnitude of the force exerted on the ball rises to a maximum value and then returns to zero when the ball leaves the bat. The time interval during which the force acts is Δt, and the magnitude of the average force is $\overline{F}$.

Linear momentum is a vector that points in the same direction as the velocity. The total linear momentum of a system of objects is the vector sum of the momenta of the individual objects.

The impulse–momentum theorem states that when a net force $\Sigma \vec{F}$ acts on an object, the impulse of the net force is equal to the change in momentum of the object:

$$(\Sigma \vec{F})\Delta t = m\vec{v}_f - m\vec{v}_0 \tag{7.4}$$

7.2 | The Principle of Conservation of Linear Momentum

External forces are those forces that agents external to the system exert on objects within the system. An isolated system is one for which the vector sum of the average external forces acting on the system is zero.

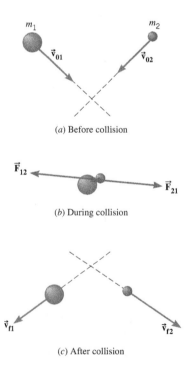

(a) Before collision

(b) During collision

(c) After collision

| FIGURE 7.6

(a) The velocities of the two objects before the collision are $\vec{v}_{01}$ and $\vec{v}_{02}$.

(b) During the collision, each object exerts a force on the other. These forces are $\vec{F}_{12}$ and $\vec{F}_{21}$.

(c) The velocities after the collision are $\vec{v}_{f1}$ and $\vec{v}_{f2}$.

The principle of conservation of linear momentum states that the total linear momentum of an isolated system remains constant. For a two-body collision, the conservation of linear momentum can be written as

$$\underbrace{m_1\vec{v}_{f1} + m_2\vec{v}_{f2}}_{\substack{\text{Final total linear} \\ \text{momentum}}} = \underbrace{m_1\vec{v}_{01} + m_2\vec{v}_{02}}_{\substack{\text{Initial total linear} \\ \text{momentum}}} \tag{7.7b}$$

where m_1 and m_2 are the masses, $\vec{v}_{f1}$ and $\vec{v}_{f2}$ are the final velocities, and $\vec{v}_{01}$ and $\vec{v}_{02}$ are the initial velocities of the objects.

7.3 | Collisions in One Dimension

An elastic collision is one in which the total kinetic energy of the system after the collision is equal to the total kinetic energy of the system before the collision.

An inelastic collision is one in which the total kinetic energy of the system is not the same before and after the collision. If the objects stick together after the collision, the collision is said to be completely inelastic.

7.4 | Collisions in Two Dimensions

When the total linear momentum is conserved in a two-dimensional collision, the x and y components of the total linear momentum are conserved separately. In the

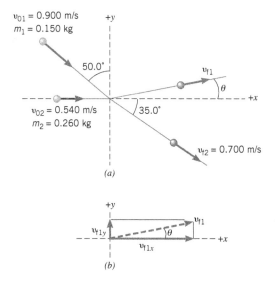

(a)

(b)

| FIGURE 7.14
(a) Top view of two balls colliding on a horizontal frictionless table.
(b) This part of the drawing shows the x and y components of the velocity of ball 1 after the collision.

case of a collision between two objects, the conservation of total linear momentum can be written as

$$\underbrace{m_1 v_{f1x} + m_2 v_{f2x}}_{\substack{x\ \text{component of final}\\ \text{total linear momentum}}} = \underbrace{m_1 v_{01x} + m_2 v_{02x}}_{\substack{x\ \text{component of initial}\\ \text{total linear momentum}}} \tag{7.9a}$$

$$\underbrace{m_1 v_{f1y} + m_2 v_{f2y}}_{\substack{y\ \text{component of final}\\ \text{total linear momentum}}} = \underbrace{m_1 v_{01y} + m_2 v_{02y}}_{\substack{y\ \text{component of initial}\\ \text{total linear momentum}}} \tag{7.9b}$$

7.5 | Center of Mass

The location of the center of mass of two particles lying on the x axis is given by

$$x_{cm} = \frac{m_1 x_1 + m_2 x_2}{m_1 + m_2} \tag{7.10}$$

where m_1 and m_2 are the masses of the particles and x_1 and x_2 are their positions relative to the coordinate origin. If the particles move with velocities v_1 and v_2, the velocity v_{cm} of the center of mass is

$$v_{cm} = \frac{m_1 v_1 + m_2 v_2}{m_1 + m_2} \tag{7.11}$$

If the total linear momentum of a system of particles remains constant during an interaction such as a collision, the velocity of the center of mass also remains constant.

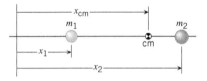

| FIGURE 7.15
The center of mass cm of the two particles is located on a line between them and lies closer to the more massive particle.

CHAPTER 8

Rotational Kinematics

8.1 | Rotational Motion and Angular Displacement

When a rigid body rotates about a fixed axis, the angular displacement is the angle swept out by a line passing through any point on the body and intersecting the axis of rotation perpendicularly. By convention, the angular displacement is positive if it is counterclockwise and negative if it is clockwise.

The radian (rad) is the SI unit of angular displacement. In radians, the angle θ is defined as the circular arc of length s traveled by a point on the rotating body divided by the radial distance r of the point from the axis:

$$\theta(\text{in radians}) = \frac{s}{r} \tag{8.1}$$

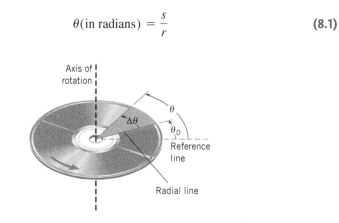

| FIGURE 8.2

The angular displacement of a CD is the angle $\Delta\theta$ swept out by a radial line as the disc turns about its axis of rotation.

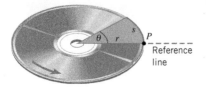

| FIGURE 8.3
In radian measure, the angle θ is defined to be the arc length s divided by the radius r.

8.2 | Angular Velocity and Angular Acceleration

The average angular velocity $\overline{\omega}$ is the angular displacement $\Delta\theta$ divided by the elapsed time Δt:

$$\overline{\omega} = \frac{\Delta\theta}{\Delta t} \tag{8.2}$$

As Δt approaches zero, the average angular velocity becomes equal to the instantaneous angular velocity ω. The magnitude of the instantaneous angular velocity is called the instantaneous angular speed.

The average angular acceleration $\overline{\alpha}$ is the change $\Delta\omega$ in the angular velocity divided by the elapsed time Δt:

$$\overline{\alpha} = \frac{\Delta\omega}{\Delta t} \tag{8.4}$$

As Δt approaches zero, the average angular acceleration becomes equal to the instantaneous angular acceleration α.

8.3 | The Equations of Rotational Kinematics

The equations of rotational kinematics apply when a rigid body rotates with a constant angular acceleration about a fixed axis. These equations relate the angular displacement $\theta - \theta_0$, the angular acceleration α, the final angular velocity ω, the initial angular velocity ω_0, and the elapsed time $t - t_0$. Assuming that $\theta_0 = 0$ rad at $t_0 = 0$ s, the equations of rotational kinematics are

$$\omega = \omega_0 + \alpha t \tag{8.4}$$

$$\theta = \tfrac{1}{2}(\omega + \omega_0)t \tag{8.6}$$

$$\theta = \omega_0 t + \tfrac{1}{2}\alpha t^2 \qquad (8.7)$$

$$\omega^2 = \omega_0^2 + 2\alpha\theta \qquad (8.8)$$

These equations may be used with any self-consistent set of units and are not restricted to radian measure.

8.4 | Angular Variables and Tangential Variables

When a rigid body rotates through an angle θ about a fixed axis, any point on the body moves on a circular arc of length s and radius r. Such a point has a tangential velocity (magnitude $= v_T$) and, possibly, a tangential acceleration (magnitude $= a_T$). The angular and tangential variables are related by the following equations:

$$s = r\theta \qquad (\theta \text{ in rad}) \qquad (8.1)$$

$$v_T = r\omega \qquad (\omega \text{ in rad/s}) \qquad (8.9)$$

$$a_T = r\alpha \qquad (\alpha \text{ in rad/s}^2) \qquad (8.10)$$

These equations refer to the magnitudes of the variables involved, without reference to positive or negative signs, and only radian measure can be used when applying them.

8.5 | Centripetal Acceleration and Tangential Acceleration

The magnitude a_c of the centripetal acceleration of a point on an object rotating with uniform or nonuniform circular motion can be expressed in terms of the radial distance r of the point from the axis and the angular speed ω:

$$a_c = r\omega^2 \qquad (\omega \text{ in rad/s}) \qquad (8.11)$$

This point experiences a total acceleration $\vec{a}$ that is the vector sum of two perpendicular acceleration components, the centripetal acceleration $\vec{a}_c$ and the tangential acceleration $\vec{a}_T$; $\vec{a} = \vec{a}_c + \vec{a}_T$.

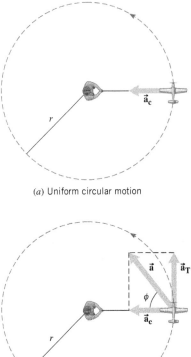

(a) Uniform circular motion

(b) Nonuniform circular motion

| **FIGURE 8.12**

(a) If a model airplane flying on a guide wire has a constant tangential speed, the motion is uniform circular motion, and the plane experiences only a centripetal acceleration $\vec{a}_c$.

(b) Nonuniform circular motion occurs when the tangential speed changes. Then there is a tangential acceleration $\vec{a}_T$ in addition to the centripetal acceleration.

8.6 | Rolling Motion

The essence of rolling motion is that there is no slipping at the point where the object touches the surface upon which it is rolling. As a result, the tangential speed v_T of a point on the outer edge of a rolling object, measured relative to the axis through the center of the object, is equal to the linear speed v with which the object moves parallel to the surface. In other words, we have

$$v = v_T = r\omega \qquad (\omega \text{ in rad/s}) \qquad (8.12)$$

The magnitudes of the tangential acceleration a_T and the linear acceleration a of a rolling object are similarly related:

$$a = a_T = r\alpha \qquad (\alpha \text{ in rad/s}^2) \qquad (8.13)$$

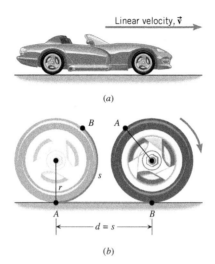

(a)

(b)

| FIGURE 8.14

(a) An automobile moves with a linear speed v.

(b) If the tires roll and do not slip, the distance d through which an axle moves equals the circular arc length s along the outer edge of a tire.

8.7 | The Vector Nature of Angular Variables

The direction of the angular velocity vector is given by a right-hand rule. Grasp the axis of rotation with your right hand, so that your fingers circle the axis in the same sense as the rotation. Your extended thumb points along the axis in the direction of the angular velocity vector.

The angular acceleration vector has the same direction as the change in the angular velocity.

CHAPTER 9

Rotational Dynamics

9.1 | The Action of Forces and Torques on Rigid Objects

The line of action of a force is an extended line that is drawn colinear with the force. The lever arm ℓ is the distance between the line of action and the axis of rotation, measured on a line that is perpendicular to both.

The torque of a force has a magnitude that is given by the magnitude F of the force times the lever arm ℓ. The torque τ is

$$\tau = F\ell \tag{9.1}$$

and is positive when the force tends to produce a counterclockwise rotation about the axis, and negative when the force tends to produce a clockwise rotation.

9.2 | Rigid Objects in Equilibrium

A rigid body is in equilibrium if it has zero translational acceleration and zero angular acceleration. In equilibrium, the net external force and the net external torque acting on the body are zero:

$$\Sigma F_x = 0 \quad \text{and} \quad \Sigma F_y = 0 \tag{4.9a and 4.9b}$$

$$\Sigma \tau = 0 \tag{9.2}$$

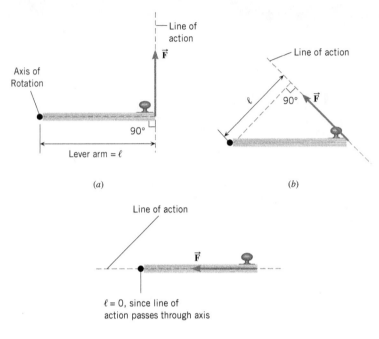

| FIGURE 9.3

In this top view, the hinges of a door appear as a black dot (•) and define the axis of rotation. The line of action and lever arm ℓ are illustrated for a force applied to the door (a) perpendicularly and (b) at an angle. (c) The lever arm is zero because the line of action passes through the axis of rotation.

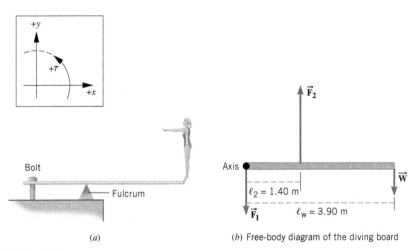

| FIGURE 9.6

(a) A diver stands at the end of a diving board.

(b) The free-body diagram for the diving board. The box at the upper left shows the positive x and y directions for the forces, as well as the positive direction (counterclockwise) for the torques.

9.3 | Center of Gravity

The center of gravity of a rigid object is the point where its entire weight can be considered to act when calculating the torque due to the weight. For a symmetrical body with uniformly distributed weight, the center of gravity is at the geometrical center of the body. When a number of objects whose weights are W_1, W_2, ... are distributed along the x axis at locations x_1, x_2, ..., the center of gravity x_{cg} is located at

$$x_{cg} = \frac{W_1 x_1 + W_2 x_2 + \cdots}{W_1 + W_2 + \cdots} \tag{9.3}$$

The center of gravity is identical to the center of mass, provided the acceleration due to gravity does not vary over the physical extent of the objects.

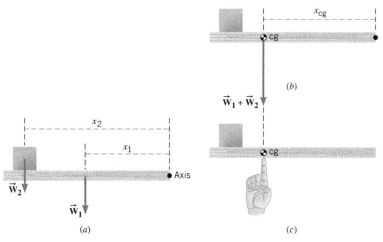

| FIGURE 9.10
(a) A box rests near the left end of a horizontal board.
(b) The total weight ($\vec{W}_1 + \vec{W}_2$) acts at the center of gravity of the group.
(c) The group can be balanced by applying an external force (due to the index finger) at the center of gravity.

9.4 | Newton's Second Law for Rotational Motion About a Fixed Axis

The moment of inertia I of a body composed of N particles is

$$I = m_1 r_1^2 + m_2 r_2^2 + \cdots + m_N r_N^2 = \Sigma m r^2 \tag{9.6}$$

where m is the mass of a particle and r is the perpendicular distance of the particle from the axis of rotation.

For a rigid body rotating about a fixed axis, Newton's second law for rotational motion is

$$\Sigma\tau = I\alpha \qquad (\alpha \text{ in rad/s}^2) \tag{9.7}$$

where $\Sigma\tau$ is the net external torque applied to the body, I is the moment of inertia of the body, and α is its angular acceleration.

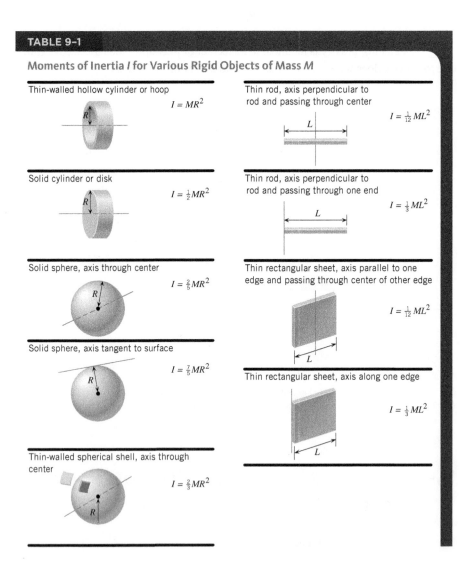

TABLE 9–1

Moments of Inertia I for Various Rigid Objects of Mass M

Thin-walled hollow cylinder or hoop
$$I = MR^2$$
R

Thin rod, axis perpendicular to rod and passing through center
$$I = \tfrac{1}{12}ML^2$$
L

Solid cylinder or disk
$$I = \tfrac{1}{2}MR^2$$
R

Thin rod, axis perpendicular to rod and passing through one end
$$I = \tfrac{1}{3}ML^2$$
L

Solid sphere, axis through center
$$I = \tfrac{2}{5}MR^2$$
R

Thin rectangular sheet, axis parallel to one edge and passing through center of other edge
$$I = \tfrac{1}{12}ML^2$$
L

Solid sphere, axis tangent to surface
$$I = \tfrac{7}{5}MR^2$$
R

Thin rectangular sheet, axis along one edge
$$I = \tfrac{1}{3}ML^2$$
L

Thin-walled spherical shell, axis through center
$$I = \tfrac{2}{3}MR^2$$
R

9.5 | Rotational Work and Energy

The rotational work W_R done by a constant torque τ in turning a rigid body through an angle θ is

$$W_R = \tau\theta \qquad (\theta \text{ in radians}) \tag{9.8}$$

The rotational kinetic energy KE_R of a rigid object rotating with an angular speed ω about a fixed axis and having a moment of inertia I is

$$KE_R = \tfrac{1}{2}I\omega^2 \qquad (\omega \text{ in rad/s}) \tag{9.9}$$

The total mechanical energy E of a rigid body is the sum of its translational kinetic energy ($\tfrac{1}{2}mv^2$), its rotational kinetic energy ($\tfrac{1}{2}I\omega^2$), and its gravitational potential energy (mgh):

$$E = \tfrac{1}{2}mv^2 + \tfrac{1}{2}I\omega^2 + mgh$$

where m is the mass of the object, v is the translational speed of its center of mass, I is its moment of inertia about an axis through the center of mass, ω is its angular speed, and h is the height of the object's center of mass relative to an arbitrary zero level.

The total mechanical energy is conserved if the net work done by external non-conservative forces and external torques is zero. When the total mechanical energy is conserved, the final total mechanical energy E_f equals the initial total mechanical energy E_0: $E_f = E_0$.

9.6 | Angular Momentum

The angular momentum of a rigid body rotating with an angular velocity ω about a fixed axis and having a moment of inertia I with respect to that axis is

$$L = I\omega \qquad (\omega \text{ in rad/s}) \tag{9.10}$$

The principle of conservation of angular momentum states that the total angular momentum of a system remains constant (is conserved) if the net average external torque acting on the system is zero. When the total angular momentum is conserved, the final angular momentum L_f equals the initial angular momentum L_0: $L_f = L_0$.

CHAPTER 10

Simple Harmonic Motion and Elasticity

10.1 | The Ideal Spring and Simple Harmonic Motion

The force that must be applied to stretch or compress an ideal spring is

$$F_x^{\text{Applied}} = kx \tag{10.1}$$

where k is the spring constant and x is the displacement of the spring from its unstrained length.

A spring exerts a restoring force on an object attached to the spring. The restoring force F_x produced by an ideal spring is

$$F_x = -kx \tag{10.2}$$

where the minus sign indicates that the restoring force points opposite to the displacement of the spring.

Simple harmonic motion is the oscillatory motion that occurs when a restoring force of the form $F_x = -kx$ acts on an object. A graphical record of position versus time for an object in simple harmonic motion is sinusoidal. The amplitude A of the motion is the maximum distance that the object moves away from its equilibrium position.

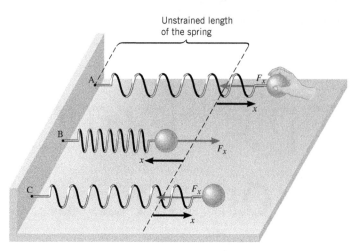

Unstrained length
of the spring

| FIGURE 10.4

The restoring force F_x (see blue arrows) produced by an ideal spring always points opposite to the displacement x (see black arrows) of the spring and leads to a back-and-forth motion of the object.

10.2 | Simple Harmonic Motion and the Reference Circle

The period T of simple harmonic motion is the time required to complete one cycle of the motion, and the frequency f is the number of cycles per second that occurs. Frequency and period are related according to

$$f = \frac{1}{T} \tag{10.5}$$

The frequency f (in Hz) is related to the angular frequency ω (in rad/s) according to

$$\omega = 2\pi f \qquad (\omega \text{ in rad/s}) \tag{10.6}$$

The maximum speed of an object in simple harmonic motion is

$$v_{\max} = A\omega \qquad (\omega \text{ in rad/s}) \tag{10.8}$$

where A is the amplitude of the motion.

The maximum acceleration of an object in simple harmonic motion is

$$a_{\max} = A\omega^2 \qquad (\omega \text{ in rad/s}) \tag{10.10}$$

The angular frequency of simple harmonic motion is

$$\omega = \sqrt{\frac{k}{m}} \qquad (\omega \text{ in rad/s}) \qquad \text{(10.11)}$$

10.3 | Energy and Simple Harmonic Motion

The elastic potential energy of an object attached to an ideal spring is

$$PE_{elastic} = \tfrac{1}{2}kx^2 \qquad \text{(10.13)}$$

The total mechanical energy E of such a system is the sum of its translational and rotational kinetic energies, gravitational potential energy, and elastic potential energy:

$$E = \tfrac{1}{2}mv^2 + \tfrac{1}{2}I\omega^2 + mgh + \tfrac{1}{2}kx^2 \qquad \text{(10.14)}$$

If external nonconservative forces like friction do no net work, the total mechanical energy of the system is conserved:

$$E_f = E_0$$

10.4 | The Pendulum

A simple pendulum is a particle of mass m attached to a frictionless pivot by a cable whose length is L and whose mass is negligible. The small-angle ($\leq 10°$) back-and-forth swinging of a simple pendulum is simple harmonic motion, but large-angle movement is not. The frequency f of the motion is given by

$$2\pi f = \sqrt{\frac{g}{L}} \qquad \text{(small angles only)} \qquad \text{(10.16)}$$

A physical pendulum consists of a rigid object, with moment of inertia I and mass m, suspended from a frictionless pivot. For small-angle displacements, the frequency f of simple harmonic motion for a physical pendulum is given by

$$2\pi f = \sqrt{\frac{mgL}{I}} \qquad \text{(small angles only)} \qquad \text{(10.15)}$$

where L is the distance between the axis of rotation and the center of gravity of the rigid object.

10.5 | Damped Harmonic Motion

Damped harmonic motion is motion in which the amplitude of oscillation decreases as time passes. Critical damping is the minimum degree of damping that eliminates any oscillations in the motion as the object returns to its equilibrium position.

10.6 | Driven Harmonic Motion and Resonance

Driven harmonic motion occurs when a driving force acts on an object along with the restoring force. Resonance is the condition under which the driving force can transmit large amounts of energy to an oscillating object, leading to large-amplitude motion. In the absence of damping, resonance occurs when the frequency of the driving force matches a natural frequency at which the object oscillates.

10.7 | Elastic Deformation

One type of elastic deformation is stretch and compression. The magnitude F of the force required to stretch or compress an object of length L_0 and cross-sectional area A by an amount ΔL is (see Figure 10.28)

$$F = Y\left(\frac{\Delta L}{L_0}\right)A \tag{10.17}$$

where Y is a constant called Young's modulus.

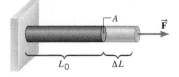

| FIGURE 10.28

In this diagram, $\vec{F}$ denotes the stretching force, A the cross-sectional area, L_0 the original length of the rod, and ΔL the amount of stretch.

TABLE 10-1

Values for the Young's Modulus of Solid Materials

Material	Young's Modulus Y (N/m²)
Aluminum	6.9×10^{10}
Bone	
Compression	9.4×10^{9}
Tension	1.6×10^{10}
Brass	9.0×10^{10}
Brick	1.4×10^{10}
Copper	1.1×10^{11}
Mohair	2.9×10^{9}
Nylon	3.7×10^{9}
Pyrex glass	6.2×10^{10}
Steel	2.0×10^{11}
Teflon	3.7×10^{8}
Titanium	1.2×10^{11}
Tungsten	3.6×10^{11}

Another type of elastic deformation is shear. The magnitude F of the shearing force required to create an amount of shear ΔX for an object of thickness L_0 and cross-sectional area A is (see Figure 10.31)

$$F = S\left(\frac{\Delta X}{L_0}\right)A \qquad (10.18)$$

where S is a constant called the shear modulus.

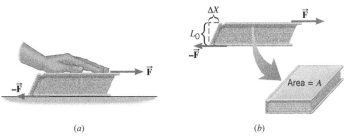

(a)   (b)

| **FIGURE 10.31**

(a) An example of a shear deformation. The shearing forces $\vec{F}$ and $-\vec{F}$ are applied parallel to the top and bottom covers of the book.

(b) The shear deformation is ΔX. The area of each cover is A, and the thickness of the book is L_0.

TABLE 10-2	
Values for the Shear Modulus of Solid Materials	
Material	Shear Modulus S (N/m²)
Aluminum	2.4×10^{10}
Bone	1.2×10^{10}
Brass	3.5×10^{10}
Copper	4.2×10^{10}
Lead	5.4×10^{9}
Nickel	7.3×10^{10}
Steel	8.1×10^{10}
Tungsten	1.5×10^{11}

A third type of elastic deformation is volume deformation, which has to do with pressure. The pressure P is the magnitude F of the force acting perpendicular to a surface divided by the area A over which the force acts:

$$P = \frac{F}{A} \tag{10.19}$$

The SI unit for pressure is N/m², a unit known as a pascal (Pa): 1 Pa $=$ 1 N/m².

The change ΔP in pressure needed to change the volume V_0 of an object by an amount ΔV is (see Figure 10.33)

$$\Delta P = -B\left(\frac{\Delta V}{V_0}\right) \tag{10.20}$$

where B is a constant known as the bulk modulus.

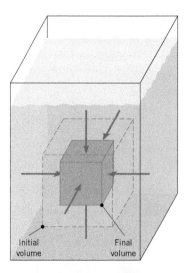

Initial volume

Final volume

| FIGURE 10.33

The arrows denote the forces that push perpendicularly on every surface of an object immersed in a liquid. The magnitude of the force per unit area is the pressure. When the pressure increases, the volume of the object decreases.

TABLE 10-3

Values for the Bulk Modulus of Solid and Liquid Materials

Material	Bulk Modulus B [N/m^2 (=Pa)]
Solids	
Aluminum	7.1×10^{10}
Brass	6.7×10^{10}
Copper	1.3×10^{11}
Diamond	4.43×10^{11}
Lead	4.2×10^{10}
Nylon	6.1×10^{9}
Osmium	4.62×10^{11}
Pyrex glass	2.6×10^{10}
Steel	1.4×10^{11}
Liquids	
Ethanol	8.9×10^{8}
Oil	1.7×10^{9}
Water	2.2×10^{9}

10.8 | Stress, Strain, and Hooke's Law

Stress is the magnitude of the force per unit area applied to an object and causes strain. For stretch/compression, the strain is the fractional change $\Delta L/L_0$ in length. For shear, the strain reflects the change in shape of the object and is given by $\Delta X/L_0$ (see Figure 10.31). For volume deformation, the strain is the fractional change in volume $\Delta V/V_0$. Hooke's law states that stress is directly proportional to strain.

CHAPTER 11

Fluids

11.1 | Mass Density

Fluids are materials that can flow, and they include gases and liquids. The mass density ρ of a substance is its mass m divided by its volume V:

$$\rho = \frac{m}{V} \tag{11.1}$$

See Table 11-1 on the next page.

The specific gravity of a substance is its mass density divided by the density of water at 4 °C (1.000×10^3 kg/m³):

$$\text{Specific gravity} = \frac{\text{Density of substance}}{1.000 \times 10^3 \text{ kg/m}^3} \tag{11.2}$$

11.2 | Pressure

The pressure P exerted by a fluid is the magnitude F of the force acting perpendicular to a surface embedded in the fluid divided by the area A over which the force acts:

$$P = \frac{F}{A} \tag{11.3}$$

The SI unit for measuring pressure is the pascal (Pa); 1 Pa = 1 N/m².

One atmosphere of pressure is 1.013×10^5 Pa or 14.7 lb/in².

TABLE 11-1

Mass Densities[a] of Common Substances

Substance	Mass Density ρ (kg/m^3)
Solids	
Aluminum	2700
Brass	8470
Concrete	2200
Copper	8890
Diamond	3520
Gold	19 300
Ice	917
Iron (steel)	7860
Lead	11 300
Quartz	2660
Silver	10 500
Wood (yellow pine)	550
Liquids	
Blood (whole, 37 °C)	1060
Ethyl alcohol	806
Mercury	13 600
Oil (hydraulic)	800
Water (4 °C)	1.000×10^3
Gases	
Air	1.29
Carbon dioxide	1.98
Helium	0.179
Hydrogen	0.0899
Nitrogen	1.25
Oxygen	1.43

[a]Unless otherwise noted, densities are given at 0 °C and 1 atm pressure.

11.3 | Pressure and Depth in a Static Fluid

In the presence of gravity, the upper layers of a fluid push downward on the layers beneath, with the result that fluid pressure is related to depth. In an incompressible static fluid whose density is ρ, the relation is

$$P_2 = P_1 + \rho g h \qquad (11.4)$$

where P_1 is the pressure at one level, P_2 is the pressure at a level that is h meters deeper, and g is the magnitude of the acceleration due to gravity.

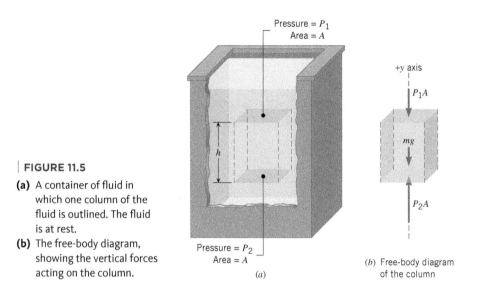

| FIGURE 11.5
(a) A container of fluid in which one column of the fluid is outlined. The fluid is at rest.
(b) The free-body diagram, showing the vertical forces acting on the column.

11.4 | Pressure Gauges

Two basic types of pressure gauges are the mercury barometer and the open-tube manometer.

The gauge pressure is the amount by which a pressure P differs from atmospheric pressure. The absolute pressure is the actual value for P.

11.5 | Pascal's Principle

Pascal's principle states that any change in the pressure applied to a completely enclosed fluid is transmitted undiminished to all parts of the fluid and the enclosing walls.

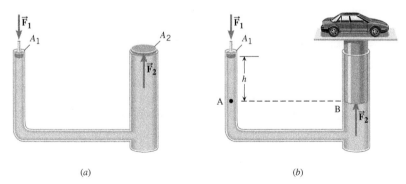

| FIGURE 11.14
(a) An external force $\vec{F}_1$ is applied to the piston on the left. As a result, a force $\vec{F}_2$ is exerted on the cap on the chamber on the right.
(b) The familiar hydraulic car lift.

11.6 | Archimedes' Principle

The buoyant force is the upward force that a fluid applies to an object that is partially or completely immersed in it.

Archimedes' principle states that the magnitude of the buoyant force equals the weight of the fluid that the partially or completely immersed object displaces:

$$\underbrace{F_B}_{\substack{\text{Magnitude of}\\ \text{buoyant force}}} = \underbrace{W_{\text{fluid}}}_{\substack{\text{Weight of}\\ \text{displaced fluid}}}$$

(11.6)

11.7 | Fluids in Motion;
11.8 | The Equation of Continuity

In steady flow, the velocity of the fluid particles at any point is constant as time passes.

An incompressible, nonviscous fluid is known as an ideal fluid.

The mass flow rate of a fluid with a density ρ, flowing with a speed v in a pipe of cross-sectional area A, is the mass per second (kg/s) flowing past a point and is given by

$$\text{Mass flow rate} = \rho A v$$

(11.7)

The equation of continuity expresses the fact that mass is conserved: what flows into one end of a pipe flows out the other end, assuming there are no additional entry or exit points in between. Expressed in terms of the mass flow rate, the equation of continuity is

$$\rho_1 A_1 v_1 = \rho_2 A_2 v_2$$

(11.8)

where the subscripts 1 and 2 denote two points along the pipe.

If a fluid is incompressible, the density at any two points is the same, $\rho_1 = \rho_2$. For an incompressible fluid, the equation of continuity becomes

$$A_1 v_1 = A_2 v_2$$

(11.9)

The product Av is known as the volume flow rate Q (in m^3/s):

$$Q = \text{Volume flow rate} = Av$$

(11.10)

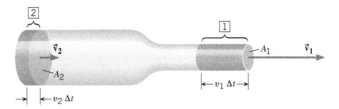

FIGURE 11.27

In general, a fluid flowing in a tube that has different cross-sectional areas A_1 and A_2 at positions 1 and 2 also has different velocities $\vec{v}_1$ and $\vec{v}_2$ at these positions.

11.9 | Bernoulli's Equation;
11.10 | Applications
of Bernoulli's Equation

In the steady flow of an ideal fluid whose density is ρ, the pressure P, the fluid speed v, and the elevation y at any two points (1 and 2) in the fluid are related by Bernoulli's equation:

$$P_1 + \tfrac{1}{2}\rho v_1{}^2 + \rho g y_1 = P_2 + \tfrac{1}{2}\rho v_2{}^2 + \rho g y_2 \qquad (11.11)$$

When the flow is horizontal ($y_1 = y_2$), Bernoulli's equation indicates that higher fluid speeds are associated with lower fluid pressures.

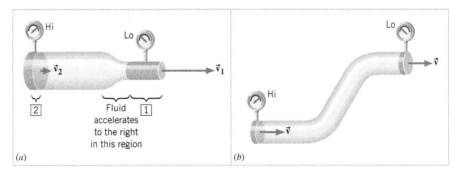

FIGURE 11.28

(a) In this horizontal pipe, the pressure in region 2 is greater than that in region 1. The difference in pressures leads to the net force that accelerates the fluid to the right.

(b) When the fluid changes elevation, the pressure at the bottom is greater than the pressure at the top, assuming that the cross-sectional area of the pipe is constant.

11.11 | Viscous Flow

The magnitude F of the tangential force required to move a fluid layer at a constant speed v, when the layer has an area A and is located a perpendicular distance y from an immobile surface, is given by

$$F = \frac{\eta A v}{y} \tag{11.13}$$

where η is the coefficient of viscosity.

A fluid whose viscosity is η, flowing through a pipe of radius R and length L, has a volume flow rate Q given by

$$Q = \frac{\pi R^4 (P_2 - P_1)}{8 \eta L} \tag{11.14}$$

where P_1 and P_2 are the pressures at the ends of the pipe.

CHAPTER 12

Temperature and Heat

12.1 | Common Temperature Scales

On the Celsius temperature scale, there are 100 equal divisions between the ice point (0 °C) and the steam point (100 °C). On the Fahrenheit temperature scale, there are 180 equal divisions between the ice point (32 °F) and the steam point (212 °F).

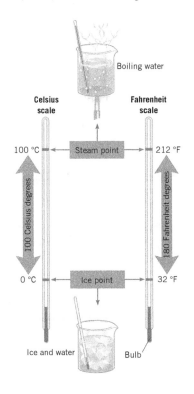

| FIGURE 12.1

The Celsius and Fahrenheit temperature scales.

12.2 | The Kelvin Temperature Scale

For scientific work, the Kelvin temperature scale is the scale of choice. One kelvin (K) is equal in size to one Celsius degree. However, the temperature T on the Kelvin scale differs from the temperature T_c on the Celsius scale by an additive constant of 273.15:

$$T = T_c + 273.15 \qquad\qquad (12.1)$$

The lower limit of temperature is called absolute zero and is designated as 0 K on the Kelvin scale.

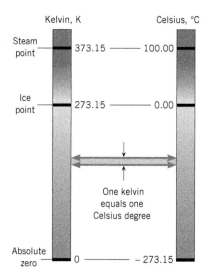

| FIGURE 12.2
A comparison of the Kelvin and Celsius temperature scales.

12.3 | Thermometers

The operation of any thermometer is based on the change in some physical property with temperature; this physical property is called a thermometric property. Examples of thermometric properties are the length of a column of mercury, electrical voltage, and electrical resistance.

12.4 | Linear Thermal Expansion

Most substances expand when heated. For linear expansion, an object of length L_0 experiences a change ΔL in length when the temperature changes by ΔT:

$$\Delta L = \alpha L_0 \, \Delta T \qquad\qquad (12.2)$$

where α is the coefficient of linear expansion.

For an object held rigidly in place, a thermal stress can occur when the object attempts to expand or contract. The stress can be large, even for small temperature changes.

When the temperature changes, a hole in a plate of solid material expands or contracts as if the hole were filled with the surrounding material.

TABLE 12-1

Coefficients of Thermal Expansion for Solids and Liquids[a]

Substance	Coefficient of Thermal Expansion $(C°)^{-1}$	
	Linear (α)	Volume (β)
Solids		
Aluminum	23×10^{-6}	69×10^{-6}
Brass	19×10^{-6}	57×10^{-6}
Concrete	12×10^{-6}	36×10^{-6}
Copper	17×10^{-6}	51×10^{-6}
Glass (common)	8.5×10^{-6}	26×10^{-6}
Glass (Pyrex)	3.3×10^{-6}	9.9×10^{-6}
Gold	14×10^{-6}	42×10^{-6}
Iron or steel	12×10^{-6}	36×10^{-6}
Lead	29×10^{-6}	87×10^{-6}
Nickel	13×10^{-6}	39×10^{-6}
Quartz (fused)	0.50×10^{-6}	1.5×10^{-6}
Silver	19×10^{-6}	57×10^{-6}
Liquids[b]		
Benzene	—	1240×10^{-6}
Carbon tetrachloride	—	1240×10^{-6}
Ethyl alcohol	—	1120×10^{-6}
Gasoline	—	950×10^{-6}
Mercury	—	182×10^{-6}
Methyl alcohol	—	1200×10^{-6}
Water	—	207×10^{-6}

[a]The values for α and β pertain to a temperature near 20 °C.
[b]Since liquids do not have fixed shapes, the coefficient of linear expansion is not defined for them.

12.5 | Volume Thermal Expansion

For volume expansion, the change ΔV in the volume of an object of volume V_0 is given by

$$\Delta V = \beta V_0 \, \Delta T \qquad\qquad (12.3)$$

where β is the coefficient of volume expansion.

When the temperature changes, a cavity in a piece of solid material expands or contracts as if the cavity were filled with the surrounding material.

12.6 | Heat and Internal Energy

The internal energy of a substance is the sum of the kinetic, potential, and other kinds of energy that the molecules of the substance have. Heat is energy that flows from a higher-temperature object to a lower-temperature object because of the difference in temperatures. The SI unit for heat is the joule (J).

12.7 | Heat and Temperature Change: Specific Heat Capacity

The heat Q that must be supplied or removed to change the temperature of a substance of mass m by an amount ΔT is

$$Q = cm\Delta T \qquad\qquad (12.4)$$

where c is a constant known as the specific heat capacity.

When materials are placed in thermal contact within a perfectly insulated container, the principle of energy conservation requires that heat lost by warmer materials equals heat gained by cooler materials.

Heat is sometimes measured with a unit called the kilocalorie (kcal). The conversion factor between kilocalories and joules is known as the mechanical equivalent of heat:

$$1 \text{ kcal} = 4186 \text{ joules}$$

12.8 | Heat and Phase Change: Latent Heat

Heat must be supplied or removed to make a material change from one phase to another. The heat Q that must be supplied or removed to change the phase of a mass m of a substance is

$$Q = mL \qquad\qquad (12.5)$$

where L is the latent heat of the substance and has SI units of J/kg. The latent heats of fusion, vaporization, and sublimation refer, respectively, to the solid/liquid, the liquid/vapor, and the solid/vapor phase changes.

TABLE 12-3

Latent Heats[a] of Fusion and Vaporization

Substance	Melting Point (°C)	Latent Heat of Fusion, L_f (J/kg)	Boiling Point (°C)	Latent Heat of Vaporization, L_v (J/kg)
Ammonia	-77.8	33.2×10^4	-33.4	13.7×10^5
Benzene	5.5	12.6×10^4	80.1	3.94×10^5
Copper	1083	20.7×10^4	2566	47.3×10^5
Ethyl alcohol	-114.4	10.8×10^4	78.3	8.55×10^5
Gold	1063	6.28×10^4	2808	17.2×10^5
Lead	327.3	2.32×10^4	1750	8.59×10^5
Mercury	-38.9	1.14×10^4	356.6	2.96×10^5
Nitrogen	-210.0	2.57×10^4	-195.8	2.00×10^5
Oxygen	-218.8	1.39×10^4	-183.0	2.13×10^5
Water	0.0	33.5×10^4	100.0	22.6×10^5

[a] The values pertain to 1 atm pressure.

12.9 | Equilibrium Between Phases of Matter

The equilibrium vapor pressure of a substance is the pressure of the vapor phase that is in equilibrium with the liquid phase. For a given substance, vapor pressure depends only on temperature. For a liquid, a plot of the equilibrium vapor pressure versus temperature is called the vapor pressure curve or vaporization curve.

The fusion curve gives the combinations of temperature and pressure for equilibrium between solid and liquid phases.

12.10 | Humidity

The relative humidity is defined as follows:

$$\text{Percent relative humidity} = \frac{\text{Partial pressure of water vapor}}{\text{Equilibrium vapor pressure of water at the existing temperature}} \times 100 \qquad (12.6)$$

The dew point is the temperature below which the water vapor in the air condenses. On the vaporization curve of water, the dew point is the temperature that corresponds to the actual pressure of water vapor in the air.

CHAPTER 13

The Transfer of Heat

13.1 | Convection

Convection is the process in which heat is carried from place to place by the bulk movement of a fluid.

During natural convection, the warmer, less dense part of a fluid is pushed upward by the buoyant force provided by the surrounding cooler and denser part.

Forced convection occurs when an external device, such as a fan or a pump, causes the fluid to move.

13.2 | Conduction

Conduction is the process whereby heat is transferred directly through a material, with any bulk motion of the material playing no role in the transfer.

Materials that conduct heat well, such as most metals, are known as thermal conductors. Materials that conduct heat poorly, such as wood, glass, and most plastics, are referred to as thermal insulators.

The heat Q conducted during a time t through a bar of length L and cross-sectional area A is

$$Q = \frac{(kA\,\Delta T)t}{L} \tag{13.1}$$

where ΔT is the temperature difference between the ends of the bar and k is the thermal conductivity of the material.

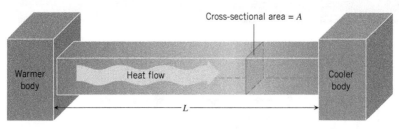

| **FIGURE 13.7**
Heat is conducted through the bar when the ends of the bar are maintained at different temperatures. The heat flows from the warmer to the cooler end.

13.3 | Radiation

Radiation is the process in which energy is transferred by means of electromagnetic waves.

All objects, regardless of their temperature, simultaneously absorb and emit electromagnetic waves. Objects that are good absorbers of radiant energy are also good emitters, and objects that are poor absorbers are also poor emitters.

An object that absorbs all the radiation incident upon it is called a perfect blackbody. A perfect blackbody, being a perfect absorber, is also a perfect emitter.

The radiant energy Q emitted during a time t by an object whose surface area is A and whose Kelvin temperature is T is given by the Stefan–Boltzmann law of radiation:

$$Q = e\sigma T^4 At \qquad (13.2)$$

where $\sigma = 5.67 \times 10^{-8}\,\text{J/(s} \cdot \text{m}^2 \cdot \text{K}^4)$ is the Stefan–Boltzmann constant and e is the emissivity, a dimensionless number characterizing the surface of the object. The emissivity lies between 0 and 1, being zero for a nonemitting surface and one for a perfect blackbody.

The net radiant power is the power an object emits minus the power it absorbs. The net radiant power P_{net} emitted by an object of temperature T located in an environment of temperature T_0 is

$$P_{\text{net}} = e\sigma A(T^4 - T_0^{\,4}) \qquad (13.3)$$

CHAPTER 14

The Ideal Gas Law and Kinetic Theory

14.1 | Molecular Mass, the Mole, and Avogadro's Number

Each element in the periodic table is assigned an atomic mass. One atomic mass unit (u) is exactly one-twelfth the mass of an atom of carbon-12. The molecular mass of a molecule is the sum of the atomic masses of its atoms.

The number of moles n contained in a sample is equal to the number of particles N (atoms or molecules) in the sample divided by the number of particles per mole N_A,

$$n = \frac{N}{N_A}$$

where N_A is called Avogadro's number and has a value of $N_A = 6.022 \times 10^{23}$ particles per mole. The number of moles is also equal to the mass m of the sample (expressed in grams) divided by the mass per mole (expressed in grams per mole):

$$n = \frac{m}{\text{Mass per mole}}$$

The mass per mole (in g/mol) of a substance has the same numerical value as the atomic or molecular mass of one of its particles (in atomic mass units).

The mass $m_{particle}$ of a particle (in grams) can be obtained by dividing the mass per mole (in g/mol) by Avogadro's number:

$$m_{particle} = \frac{\text{Mass per mole}}{N_A}$$

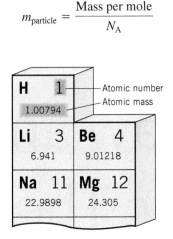

| FIGURE 14.1
A portion of the periodic table showing the atomic number and atomic mass of each element. In the periodic table it is customary to omit the symbol "u" denoting the atomic mass unit.

14.2 | The Ideal Gas Law

The ideal gas law relates the absolute pressure P, the volume V, the number n of moles, and the Kelvin temperature T of an ideal gas according to

$$PV = nRT \qquad\qquad (14.1)$$

where $R = 8.31$ J/(mol $\cdot$ K) is the universal gas constant. An alternative form of the ideal gas law is

$$PV = NkT \qquad\qquad (14.2)$$

where N is the number of particles and $k = \dfrac{R}{N_A}$ is Boltzmann's constant. A real gas behaves as an ideal gas when its density is low enough that its particles do not interact, except via elastic collisions.

A form of the ideal gas law that applies when the number of moles and the temperature are constant is known as Boyle's law. Using the subscripts "i" and "f" to denote, respectively, initial and final conditions, we can write Boyle's law as

$$P_iV_i = P_fV_f \qquad\qquad (14.3)$$

A form of the ideal gas law that applies when the number of moles and the pressure are constant is called Charles' law:

$$\frac{V_i}{T_i} = \frac{V_f}{T_f} \qquad (14.4)$$

14.3 | Kinetic Theory of Gases

The distribution of particle speeds in an ideal gas at constant temperature is the Maxwell speed distribution (see Figure 14.8). The kinetic theory of gases indicates that the Kelvin temperature T of an ideal gas is related to the average translational kinetic energy $\overline{\text{KE}}$ of a particle according to

$$\overline{\text{KE}} = \tfrac{1}{2}mv_{\text{rms}}^2 = \tfrac{3}{2}kT \qquad (14.6)$$

where v_{rms} is the root-mean-square speed of the particles.

The internal energy U of n moles of a monatomic ideal gas is

$$U = \tfrac{3}{2}nRT \qquad (14.7)$$

The internal energy of any type of ideal gas (e.g., monatomic, diatomic) is proportional to its Kelvin temperature.

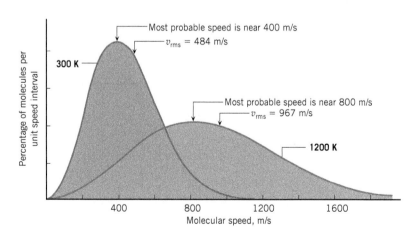

| FIGURE 14.8

The Maxwell distribution curves for molecular speeds in oxygen gas at temperatures of 300 and 1200 K.

14.4 | Diffusion

Diffusion is the process whereby solute molecules move through a solvent from a region of higher solute concentration to a region of lower solute concentration. Fick's law of diffusion states that the mass m of solute that diffuses in a time t through the solvent in a channel of length L and cross-sectional area A is

$$m = \frac{(DA\ \Delta C)t}{L} \qquad\qquad (14.8)$$

where ΔC is the solute concentration difference between the ends of the channel and D is the diffusion constant.

CHAPTER 15

Thermodynamics

15.1 | Thermodynamic Systems and Their Surroundings

A thermodynamic system is the collection of objects on which attention is being focused, and the surroundings are everything else in the environment. The state of a system is the physical condition of the system, as described by values for physical parameters, often pressure, volume, and temperature.

15.2 | The Zeroth Law of Thermodynamics

Two systems are in thermal equilibrium if there is no net flow of heat between them when they are brought into thermal contact.

Temperature is the indicator of thermal equilibrium in the sense that there is no net flow of heat between two systems in thermal contact that have the same temperature.

The zeroth law of thermodynamics states that two systems individually in thermal equilibrium with a third system are in thermal equilibrium with each other.

15.3 | The First Law of Thermodynamics

The first law of thermodynamics states that due to heat Q and work W, the internal energy of a system changes from its initial value of U_i to a final value of U_f according to

$$\Delta U = U_f - U_i = Q - W \qquad (15.1)$$

Q is positive when the system gains heat and negative when it loses heat. W is positive when work is done by the system and negative when work is done on the system.

The first law of thermodynamics is the conservation-of-energy principle applied to heat, work, and the change in the internal energy.

The internal energy is called a function of state because it depends only on the state of the system and not on the method by which the system came to be in a given state.

15.4 | Thermal Processes

A thermal process is quasi-static when it occurs slowly enough that a uniform pressure and temperature exist throughout the system at all times.

An isobaric process is one that occurs at constant pressure. The work W done when a system changes at a constant pressure P from an initial volume V_i to a final volume V_f is

$$W = P\,\Delta V = P(V_f - V_i) \tag{15.2}$$

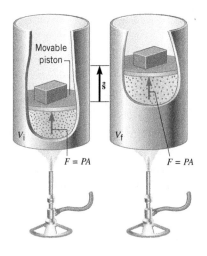

| FIGURE 15.4

The substance in the chamber is expanding isobarically because the pressure is held constant by the external atmosphere and the weight of the piston and the block.

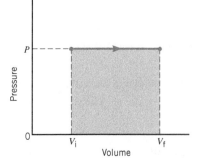

| FIGURE 15.5

For an isobaric process, a pressure-versus-volume plot is a horizontal straight line, and the work done $[W = P(V_f - V_i)]$ is the colored rectangular area under the graph.

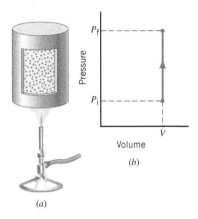

| FIGURE 15.6

(a) The substance in the chamber is being heated isochorically because the rigid chamber keeps the volume constant.

(b) The pressure–volume plot for an isochoric process is a vertical straight line. The area under the graph is zero, indicating that no work is done.

An isochoric process is one that takes place at constant volume, and no work is done in such a process.

An isothermal process is one that takes place at constant temperature.

An adiabatic process is one that takes place without the transfer of heat.

The work done in any kind of quasi-static process is given by the area under the corresponding pressure-versus-volume graph.

15.5 | Thermal Processes Using an Ideal Gas

When n moles of an ideal gas change quasi-statically from an initial volume V_i to a final volume V_f at a constant Kelvin temperature T, the work done is

$$W = nRT \ln\left(\frac{V_f}{V_i}\right) \qquad (15.3)$$

When n moles of a monatomic ideal gas change quasi-statically and adiabatically from an initial temperature T_i to a final temperature T_f, the work done is

$$W = \tfrac{3}{2}nR(T_i - T_f) \qquad (15.4)$$

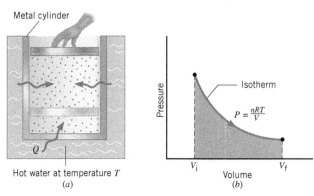

| FIGURE 15.8

(a) The ideal gas in the cylinder is expanding isothermally at temperature T. The force holding the piston in place is reduced slowly, so the expansion occurs quasi-statically.

(b) The work done by the gas is given by the colored area.

During an adiabatic process, and in addition to the ideal gas law, an ideal gas obeys the relation

$$P_i V_i{}^\gamma = P_f V_f{}^\gamma \tag{15.5}$$

where $\gamma = c_P/c_V$ is the ratio of the specific heat capacities at constant pressure and constant volume.

TABLE 15-1

Summary of Thermal Processes

Type of Thermal Process	Work Done	First Law of Thermodynamics ($\Delta U = Q - W$)
Isobaric (constant pressure)	$W = P(V_f - V_i)$	$\Delta U = Q - \underbrace{P(V_f - V_i)}_{W}$
Isochoric (constant volume)	$W = 0\ \text{J}$	$\Delta U = Q - \underbrace{0\ \text{J}}_{W}$
Isothermal (constant temperature)	$W = nRT \ln\left(\dfrac{V_f}{V_i}\right)$ (for an ideal gas)	$\underbrace{0\ \text{J}}_{\Delta U\ \text{for an ideal gas}} = Q - \underbrace{nRT \ln\left(\dfrac{V_f}{V_i}\right)}_{W}$
Adiabatic (no heat flow)	$W = \frac{3}{2}nR(T_i - T_f)$ (for a monatomic ideal gas)	$\Delta U = \underbrace{0\ \text{J}}_{Q} - \underbrace{\frac{3}{2}nR(T_i - T_f)}_{W}$

15.6 | Specific Heat Capacities

The molar specific heat capacity C of a substance determines how much heat Q is added or removed when the temperature of n moles of the substance changes by an amount ΔT:

$$Q = Cn\Delta T \tag{15.6}$$

For a monatomic ideal gas, the molar specific heat capacities at constant pressure and constant volume are, respectively,

$$C_P = \tfrac{5}{2}R \tag{15.7}$$

$$C_V = \tfrac{3}{2}R \tag{15.8}$$

where R is the ideal gas constant. For a diatomic ideal gas at moderate temperatures that do not allow vibration to occur, these values are $C_P = \tfrac{7}{2}R$ and $C_V = \tfrac{5}{2}R$. For any type of ideal gas, the difference between C_P and C_V is

$$C_P - C_V = R \tag{15.10}$$

15.7 | The Second Law of Thermodynamics

The second law of thermodynamics can be stated in a number of equivalent forms. In terms of heat flow, the second law declares that heat flows spontaneously from a substance at a higher temperature to a substance at a lower temperature and does not flow spontaneously in the reverse direction.

15.8 | Heat Engines

A heat engine produces work (magnitude $= |W|$) from input heat (magnitude $= |Q_H|$) that is extracted from a heat reservoir at a relatively high temperature. The engine rejects heat (magnitude $= |Q_C|$) into a reservoir at a relatively low temperature.

The efficiency e of a heat engine is

$$e = \frac{\text{Work done}}{\text{Input heat}} = \frac{|W|}{|Q_H|} \tag{15.11}$$

The conservation of energy requires that $|Q_H|$ must be equal to $|W|$ plus $|Q_C|$:

$$|Q_H| = |W| + |Q_C| \qquad (15.12)$$

By combining Equation 15.12 with Equation 15.11, the efficiency of a heat engine can also be written as

$$e = 1 - \frac{|Q_C|}{|Q_H|} \qquad (15.13)$$

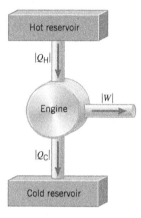

| FIGURE 15.10
This schematic representation of a heat engine shows the input heat (magnitude $= |Q_H|$) that originates from the hot reservoir, the work (magnitude $= |W|$) that the engine does, and the heat (magnitude $= |Q_C|$) that the engine rejects to the cold reservoir.

15.9 | Carnot's Principle and the Carnot Engine

A reversible process is one in which *both* the system and its environment can be returned to exactly the states they were in before the process occurred.

Carnot's principle is an alternative statement of the second law of thermodynamics. It states that no irreversible engine operating between two reservoirs at constant temperatures can have a greater efficiency than a reversible engine operating between the same temperatures. Furthermore, all reversible engines operating between the same temperatures have the same efficiency.

A Carnot engine is a reversible engine in which all input heat (magnitude $= |Q_H|$) originates from a hot reservoir at a single Kelvin temperature T_H and all rejected heat (magnitude $= |Q_C|$) goes into a cold reservoir at a single Kelvin temperature T_C. For a Carnot engine

$$\frac{|Q_C|}{|Q_H|} = \frac{T_C}{T_H} \tag{15.14}$$

The efficiency e_{Carnot} of a Carnot engine is the maximum efficiency that an engine operating between two fixed temperatures can have:

$$e_{Carnot} = 1 - \frac{T_C}{T_H} \tag{15.15}$$

15.10 | Refrigerators, Air Conditioners, and Heat Pumps

Refrigerators, air conditioners, and heat pumps are devices that utilize work (magnitude $= |W|$) to make heat (magnitude $= |Q_C|$) flow from a lower Kelvin temperature T_C to a higher Kelvin temperature T_H. In the process (the refrigeration process) they deposit heat (magnitude $= |Q_H|$) at the higher temperature. The principle of the conservation of energy requires that $|Q_H| = |W| + |Q_C|$.

If the refrigeration process is ideal, in the sense that it occurs reversibly, the devices are called Carnot devices and the relation $|Q_C|/|Q_H| = T_C/T_H$ (Equation 15.14) holds.

The coefficient of performance of a refrigerator or an air conditioner is

$$\text{Coefficient of performance} = \frac{|Q_C|}{|W|} \tag{15.16}$$

The coefficient of performance of a heat pump is

$$\text{Coefficient of performance} = \frac{|Q_H|}{|W|} \tag{15.17}$$

15.11 | Entropy

The change in entropy ΔS for a process in which heat Q enters or leaves a system reversibly at a constant Kelvin temperature T is

$$\Delta S = \left(\frac{Q}{T}\right)_R \tag{15.18}$$

where the subscript R stands for "reversible."

The second law of thermodynamics can be stated in a number of equivalent forms. In terms of entropy, the second law states that the total entropy of the universe does not change when a reversible process occurs ($\Delta S_{universe} = 0$ J/K) and increases when an irreversible process occurs ($\Delta S_{universe} > 0$ J/K).

Irreversible processes cause energy to be degraded in the sense that part of the energy becomes unavailable for the performance of work. The energy $W_{unavailable}$ that is unavailable for doing work because of an irreversible process is

$$W_{unavailable} = T_0 \, \Delta S_{universe} \qquad\qquad (15.19)$$

where $\Delta S_{universe}$ is the total entropy change of the universe and T_0 is the Kelvin temperature of the coldest reservoir into which heat can be rejected.

Increased entropy is associated with a greater degree of disorder and decreased entropy with a lesser degree of disorder (more order).

15.12 | The Third Law of Thermodynamics

The third law of thermodynamics states that it is not possible to lower the temperature of any system to absolute zero ($T = 0$ K) in a finite number of steps.

CHAPTER 16

Waves and Sound

16.1 | The Nature of Waves

A wave is a traveling disturbance and carries energy from place to place. In a transverse wave, the disturbance occurs perpendicular to the direction of travel of the wave. In a longitudinal wave, the disturbance occurs parallel to the line along which the wave travels.

16.2 | Periodic Waves

A periodic wave consists of cycles or patterns that are produced over and over again by the source of the wave. The amplitude of the wave is the maximum excursion of a particle of the medium from the particle's undisturbed position. The wavelength λ is the distance along the length of the wave between two successive equivalent points, such as two crests or two troughs. The period T is the time required for the wave to travel a distance of one wavelength. The frequency f (in hertz) is the number of wave cycles per second that passes an observer and is the reciprocal of the period (in seconds):

$$f = \frac{1}{T} \tag{10.5}$$

The speed v of a wave is related to its wavelength and frequency according to

$$v = f\lambda \tag{16.1}$$

See Figure 16.5 on the next page.

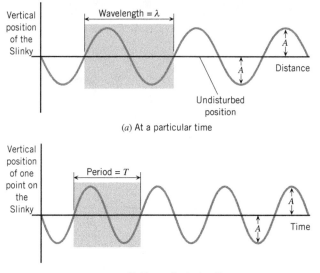

(a) At a particular time

(b) At a particular location

| FIGURE 16.5

A transverse wave on a Slinky. One cycle of the wave is shaded in color, and the amplitude of the wave is denoted as A.

16.3 | The Speed of a Wave on a String

The speed of a wave depends on the properties of the medium in which the wave travels. For a transverse wave on a string that has a tension F and a mass per unit length m/L, the wave speed is

$$v = \sqrt{\frac{F}{m/L}} \qquad (16.2)$$

The mass per unit length is also called the linear density.

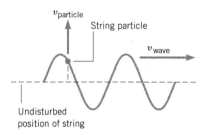

| FIGURE 16.9

A transverse wave on a string is moving to the right with a constant speed v_{wave}. A string particle moves up and down in simple harmonic motion about the undisturbed position of the string. A string particle moves with a speed v_{particle}.

16.4 | The Mathematical Description of a Wave

When a wave of amplitude A, frequency f, and wavelength λ moves in the $+x$ direction through a medium, the wave causes a displacement y of a particle at position x according to

$$y = A \sin\left(2\pi ft - \frac{2\pi x}{\lambda}\right) \tag{16.3}$$

For a wave moving in the $-x$ direction, the expression is

$$y = A \sin\left(2\pi ft + \frac{2\pi x}{\lambda}\right) \tag{16.4}$$

16.5 | The Nature of Sound

Sound is a longitudinal wave that can be created only in a medium; it cannot exist in a vacuum. Each cycle of a sound wave includes one condensation (a region of greater than normal pressure) and one rarefaction (a region of less than normal pressure).

A sound wave with a single frequency is called a pure tone. Frequencies less than 20 Hz are called infrasonic. Frequencies greater than 20 kHz are called ultrasonic. The brain interprets the frequency detected by the ear primarily in terms of the subjective quality known as pitch. A high-pitched sound is one with a large frequency (e.g., piccolo). A low-pitched sound is one with a small frequency (e.g., tuba).

The pressure amplitude of a sound wave is the magnitude of the maximum change in pressure, measured relative to the undisturbed pressure. The pressure amplitude is associated with the subjective quality of loudness. The larger the pressure amplitude, the louder the sound.

16.6 | The Speed of Sound

The speed of sound v depends on the properties of the medium. In an ideal gas, the speed of sound is

$$v = \sqrt{\frac{\gamma kT}{m}} \tag{16.5}$$

where $\gamma = c_P/c_V$ is the ratio of the specific heat capacities at constant pressure and constant volume, k is Boltzmann's constant, T is the Kelvin temperature, and m is the mass of a molecule of the gas. In a liquid, the speed of sound is

$$v = \sqrt{\frac{B_{ad}}{\rho}} \qquad (16.6)$$

where B_{ad} is the adiabatic bulk modulus and ρ is the mass density. In a solid that has a Young's modulus of Y and the shape of a long slender bar, the speed of sound is

$$v = \sqrt{\frac{Y}{\rho}} \qquad (16.7)$$

TABLE 16-1

Speed of Sound in Gases, Liquids, and Solids

Substance	Speed (m/s)
Gases	
Air (0 °C)	331
Air (20 °C)	343
Carbon dioxide (0 °C)	259
Oxygen (0 °C)	316
Helium (0 °C)	965
Liquids	
Chloroform (20 °C)	1004
Ethyl alcohol (20 °C)	1162
Mercury (20 °C)	1450
Fresh water (20 °C)	1482
Seawater (20 °C)	1522
Solids	
Copper	5010
Glass (Pyrex)	5640
Lead	1960
Steel	5960

16.7 | Sound Intensity

The intensity I of a sound wave is the power P that passes perpendicularly through a surface divided by the area A of the surface

$$I = \frac{P}{A} \qquad (16.8)$$

The SI unit for intensity is watts per square meter (W/m²). The smallest sound intensity that the human ear can detect is known as the threshold of hearing and is about 1×10^{-12} W/m² for a 1-kHz sound. When a source radiates sound uniformly in all directions and no reflections are present, the intensity of the sound is inversely proportional to the square of the distance from the source, according to

$$I = \frac{P}{4\pi r^2} \tag{16.9}$$

16.8 | Decibels

The intensity level β (in decibels) is used to compare a sound intensity I to the sound intensity I_0 of a reference level:

$$\beta = (10 \text{ dB}) \log\left(\frac{I}{I_0}\right) \tag{16.10}$$

The decibel, like the radian, is dimensionless. An intensity level of zero decibels means that $I = I_0$. One decibel is approximately the smallest change in loudness that an average listener with healthy hearing can detect. An increase of ten decibels in the intensity level corresponds approximately to a doubling of the loudness of the sound.

16.9 | The Doppler Effect

The Doppler effect is the change in frequency detected by an observer because the sound source and the observer have different velocities with respect to the medium of sound propagation. If the observer and source move with speeds v_o and v_s, respectively, and if the medium is stationary, the frequency f_o detected by the observer is

$$f_o = f_s \left(\frac{1 \pm \dfrac{v_o}{v}}{1 \mp \dfrac{v_s}{v}} \right) \tag{16.15}$$

where f_s is the frequency of the sound emitted by the source and v is the speed of sound. In the numerator, the plus sign applies when the observer moves toward the source, and the minus sign applies when the observer moves away from the source. In the denominator, the minus sign is used when the source moves toward the observer, and the plus sign is used when the source moves away from the observer. See Figure 16.27 on the next page.

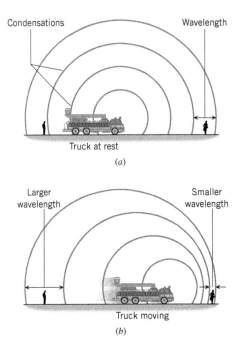

| FIGURE 16.27

(a) When the truck is stationary, the wavelength of the sound is the same in front of and behind the truck.

(b) When the truck is moving, the wavelength in front of the truck becomes smaller, while the wavelength behind the truck becomes larger.

CHAPTER 17

The Principle of Linear Superposition and Interference Phenomena

17.1 | The Principle of Linear Superposition

The principle of linear superposition states that when two or more waves are present simultaneously at the same place, the resultant disturbance is the sum of the disturbances from the individual waves.

17.2 | Constructive and Destructive Interference of Sound Waves

Constructive interference occurs at a point when two waves meet there crest-to-crest and trough-to-trough, thus reinforcing each other. Destructive interference occurs when the waves meet crest-to-trough and cancel each other.

When waves meet crest-to-crest and trough-to-trough, they are exactly in phase. When they meet crest-to-trough, they are exactly out of phase.

For two wave sources vibrating in phase, a difference in path lengths that is zero or an integer number (1, 2, 3, . . .) of wavelengths leads to constructive interference; a difference in path lengths that is a half-integer number ($\frac{1}{2}$, $1\frac{1}{2}$, $2\frac{1}{2}$, . . .) of wavelengths leads to destructive interference.

For two wave sources vibrating out of phase, a difference in path lengths that is a half-integer number ($\frac{1}{2}$, $1\frac{1}{2}$, $2\frac{1}{2}$, . . .) of wavelengths leads to constructive interference; a difference in path lengths that is zero or an integer number (1, 2, 3, . . .) of wavelengths leads to destructive interference.

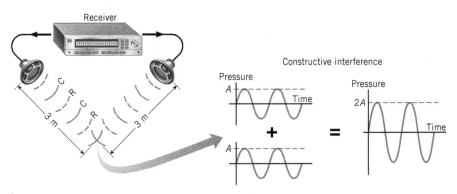

| **FIGURE 17.3**
As a result of constructive interference between the two sound waves (amplitude = A), a loud sound (amplitude = 2A) is heard at an overlap point located equally distant from two in-phase speakers (C, condensation; R, rarefaction).

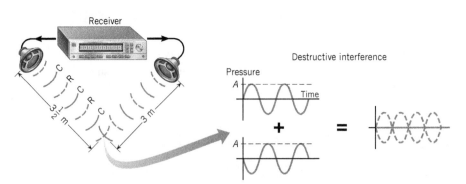

| **FIGURE 17.4**
The speakers in this drawing vibrate in phase. However, the left speaker is one-half of a wavelength ($\frac{1}{2}$m) farther from the overlap point than the right speaker. Because of destructive interference, no sound is heard at the overlap point (C, condensation; R, rarefaction).

17.3 | Diffraction

Diffraction is the bending of a wave around an obstacle or the edges of an opening. The angle through which the wave bends depends on the ratio of the wavelength λ of the wave to the width D of the opening; the greater the ratio λ/D, the greater the angle.

When a sound wave of wavelength λ passes through an opening, the first place where the intensity of the sound is a minimum relative to the center of the opening is specified by the angle θ. If the opening is a rectangular slit of width D, such as a doorway, the angle is

$$\sin \theta = \frac{\lambda}{D} \qquad (17.1)$$

If the opening is a circular opening of diameter D, such as that in a loudspeaker, the angle is

$$\sin \theta = 1.22 \frac{\lambda}{D} \qquad (17.2)$$

17.4 | Beats

Beats are the periodic variations in amplitude that arise from the linear superposition of two waves that have slightly different frequencies. When the waves are sound waves, the variations in amplitude cause the loudness to vary at the beat frequency, which is the difference between the frequencies of the waves.

17.5 | Transverse Standing Waves

A standing wave is the pattern of disturbance that results when oppositely traveling waves of the same frequency and amplitude pass through each other. A standing wave has places of minimum and maximum vibration called, respectively, nodes and antinodes.

Under resonance conditions, standing waves can be established only at certain natural frequencies. The frequencies in this series (f_1, $2f_1$, $3f_1$, etc.) are called harmonics. The lowest frequency f_1 is called the first harmonic, the next frequency $2f_1$ is the second harmonic, and so on.

For a string that is fixed at both ends and has a length L, the natural frequencies are

$$f_n = n\left(\frac{v}{2L}\right) \qquad n = 1, 2, 3, 4, \ldots \qquad (17.3)$$

where v is the speed of the wave on the string and n is a positive integer.

17.6 | Longitudinal Standing Waves

For a gas in a cylindrical tube open at both ends, the natural frequencies of vibration are

$$f_n = n\left(\frac{v}{2L}\right) \qquad n = 1, 2, 3, 4, \ldots \qquad \textbf{(17.4)}$$

where v is the speed of sound in the gas and L is the length of the tube.

For a gas in a cylindrical tube open at only one end, the natural frequencies of vibration are

$$f_n = n\left(\frac{v}{4L}\right) \qquad n = 1, 3, 5, 7, \ldots \qquad \textbf{(17.5)}$$

17.7 | Complex Sound Waves

A complex sound wave consists of a mixture of a fundamental frequency and overtone frequencies.

CHAPTER 18

Electric Forces and Electric Fields

18.1 | The Origin of Electricity

There are two kinds of electric charge: positive and negative. The SI unit of electric charge is the coulomb (C). The magnitude of the charge on an electron or a proton is

$$e = 1.60 \times 10^{-19}\,\text{C}$$

Since the symbol e denotes a magnitude, it has no algebraic sign. Thus, the electron carries a charge of $-e$, and the proton carries a charge of $+e$.

The charge on any object, whether positive or negative, is quantized, in the sense that the charge consists of an integer number of protons or electrons.

18.2 | Charged Objects and the Electric Force

The law of conservation of electric charge states that the net electric charge of an isolated system remains constant during any process.

Like charges repel and unlike charges attract each other.

See Figure 18.3 on the next page.

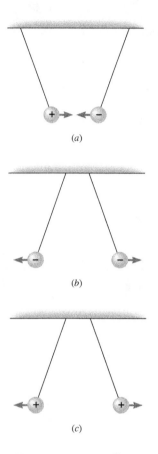

| **FIGURE 18.3**

(a) A positive charge (+) and a negative charge (−) attract each other.
(b) Two negative charges repel each other.
(c) Two positive charges repel each other.

18.3 | Conductors and Insulators

An electrical conductor is a material, such as copper, that conducts electric charge readily.

An electrical insulator is a material, such as rubber, that conducts electric charge poorly.

18.4 | Charging by Contact and by Induction

Charging by contact is the process of giving one object a net electric charge by placing it in contact with an object that is already charged.

Charging by induction is the process of giving an object a net electric charge without touching it to a charged object.

18.5 | Coulomb's Law

A point charge is a charge that occupies so little space that it can be regarded as a mathematical point.

Coulomb's law gives the magnitude F of the electric force that two point charges q_1 and q_2 exert on each other:

$$F = k\frac{|q_1||q_2|}{r^2} \tag{18.1}$$

where $|q_1|$ and $|q_2|$ are the magnitudes of the charges and have no algebraic sign. The term k is a constant and has the value $k = 8.99 \times 10^9$ N $\cdot$ m²/C². The force specified by Equation 18.1 acts along the line between the two charges.

The permittivity of free space ϵ_0 is defined by the relation

$$k = \frac{1}{4\pi\epsilon_0}$$

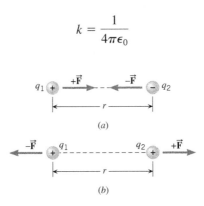

(a)

(b)

| FIGURE 18.9
Each point charge exerts a force on the other. Regardless of whether the forces are (a) attractive or (b) repulsive, they are directed along the line between the charges and have equal magnitudes.

18.6 | The Electric Field

The electric field $\vec{E}$ at a given spot is a vector and is the electrostatic force $\vec{F}$ experienced by a very small test charge q_0 placed at that spot divided by the charge itself:

$$\vec{E} = \frac{\vec{F}}{q_0} \tag{18.2}$$

The direction of the electric field is the same as the direction of the force on a positive test charge. The SI unit for the electric field is the newton per coulomb (N/C). The source of the electric field at any spot is the charged objects surrounding that spot.

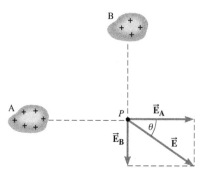

| FIGURE 18.16
The electric field contributions $\vec{E}_A$ and $\vec{E}_B$, which come from the two charge distributions, are added vectorially to obtain the net field $\vec{E}$ at point P.

The magnitude of the electric field created by a point charge q is

$$E = \frac{k|q|}{r^2} \tag{18.3}$$

where $|q|$ is the magnitude of the charge and has no algebraic sign and r is the distance from the charge. The electric field $\vec{E}$ points away from a positive charge and toward a negative charge.

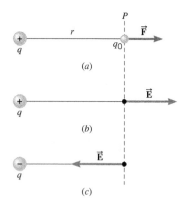

| FIGURE 18.18
(a) At location P, a positive test charge q_0 experiences a repulsive force $\vec{F}$ due to the positive point charge q.
(b) At P, the electric field $\vec{E}$ is directed to the right.
(c) If the charge q were negative rather than positive, the electric field would have the same magnitude as in (b) but would point to the left.

For a parallel plate capacitor that has a charge per unit area of σ on each plate, the magnitude of the electric field between the plates is

$$E = \frac{\sigma}{\epsilon_0} \qquad (18.4)$$

18.7 | Electric Field Lines

Electric field lines are lines that can be thought of as a "map," insofar as the lines provide information about the direction and strength of the electric field. The lines are directed away from positive charges and toward negative charges. The direction of the lines gives the direction of the electric field, since the electric field vector at a point is tangent to the line at that point. The electric field is strongest in regions where the number of lines per unit area passing perpendicularly through a surface is the greatest—that is, where the lines are packed together most tightly.

18.8 | The Electric Field Inside a Conductor: Shielding

Excess negative or positive charge resides on the surface of a conductor at equilibrium under electrostatic conditions. In such a situation, the electric field at any point within the conducting material is zero, and the electric field just outside the surface of the conductor is perpendicular to the surface.

18.9 | Gauss' Law

The electric flux Φ_E through a surface is related to the magnitude E of the electric field, the area A of the surface, and the angle ϕ that specifies the direction of the field relative to the normal to the surface:

$$\Phi_E = \Sigma (E \cos \phi) \Delta A \qquad (18.6)$$

Gauss' law states that the electric flux through a closed surface (a Gaussian surface) is equal to the net charge Q enclosed by the surface divided by ϵ_0, the permittivity of free space:

$$\Phi_E = \Sigma (E \cos \phi) \Delta A = \frac{Q}{\epsilon_0} \qquad (18.7)$$

See Figure 18.32 on the next page.

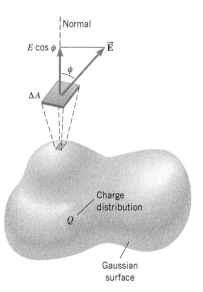

| **FIGURE 18.32**

The charge distribution Q is surrounded by an arbitrarily shaped Gaussian surface. The electric flux Φ through any tiny segment of the surface is the product of $E \cos \phi$ and the area ΔA of the segment: $\Phi = (E \cos \phi) \Delta A$. The angle ϕ is the angle between the electric field and the normal to the surface.

CHAPTER 19

Electric Potential Energy and the Electric Potential

19.1 | Potential Energy

When a positive test charge $+q_0$ moves from point A to point B in an electric field, work W_{AB} is done by the electric force. The work equals the electric potential energy (EPE) at A minus that at B:

$$W_{AB} = \text{EPE}_A - \text{EPE}_B \qquad (19.1)$$

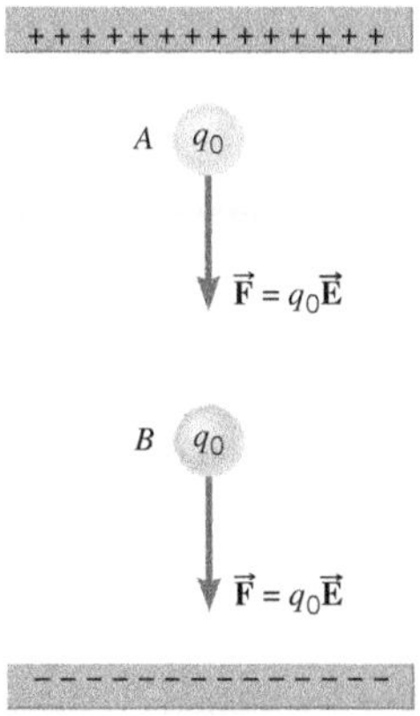

| FIGURE 19.2

Because of the electric field $\vec{E}$, an electric force, $\vec{F} = q_0\vec{E}$, is exerted on a positive test charge $+q_0$. Work is done by the force as the charge moves from A to B.

The electric force is a conservative force, so the path along which the test charge moves from A to B is of no consequence, for the work W_{AB} is the same for all paths.

19.2 | The Electric Potential Difference

The electric potential V at a given point is the electric potential energy of a small test charge q_0 situated at that point divided by the charge itself:

$$V = \frac{\text{EPE}}{q_0} \tag{19.3}$$

The SI unit of electric potential is the joule per coulomb (J/C), or volt (V). The electric potential difference between two points A and B is

$$V_B - V_A = \frac{\text{EPE}_B}{q_0} - \frac{\text{EPE}_A}{q_0} = \frac{-W_{AB}}{q_0} \tag{19.4}$$

The potential difference between two points (or between two equipotential surfaces) is often called the "voltage."

A positive charge accelerates from a region of higher potential toward a region of lower potential. Conversely, a negative charge accelerates from a region of lower potential toward a region of higher potential.

An electron volt (eV) is a unit of energy. The relationship between electron volts and joules is $1 \text{ eV} = 1.60 \times 10^{-19}$ J.

The total energy E of a system is the sum of its translational $(\frac{1}{2}mv^2)$ and rotational $(\frac{1}{2}I\omega^2)$ kinetic energies, gravitational potential energy (mgh), elastic potential energy $(\frac{1}{2}kx^2)$, and electric potential energy (EPE):

$$E = \tfrac{1}{2}mv^2 + \tfrac{1}{2}I\omega^2 + mgh + \tfrac{1}{2}kx^2 + \text{EPE}$$

If external nonconservative forces like friction do no net work, the total energy of the system is conserved. That is, the final total energy E_f is equal to the initial total energy E_0; $E_f = E_0$.

19.3 | The Electric Potential Difference Created by Point Charges

The electric potential V at a distance r from a point charge q is

$$V = \frac{kq}{r} \tag{19.6}$$

where $k = 8.99 \times 10^9$ N $\cdot$ m^2/C^2. This expression for V assumes that the electric potential is zero at an infinite distance away from the charge.

The total electric potential at a given location due to two or more charges is the algebraic sum of the potentials due to each charge.

The total potential energy of a group of charges is the amount by which the electric potential energy of the group differs from its initial value when the charges are infinitely far apart and far away. It is also equal to the work required to assemble the group, one charge at a time, starting with the charges infinitely far apart and far away.

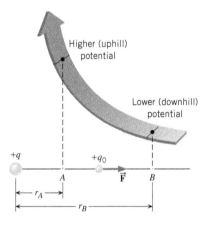

| FIGURE 19.7
The positive test charge $+q_0$ experiences a repulsive force $\vec{F}$ due to the positive point charge $+q$. As a result, work is done by this force when the test charge moves from A to B. Consequently, the electric potential is higher (uphill) at A and lower (downhill) at B.

19.4 | Equipotential Surfaces and Their Relation to the Electric Field

An equipotential surface is a surface on which the electric potential is the same everywhere. The electric force does no work as a charge moves on an equipotential surface, because the force is always perpendicular to the displacement of the charge.

The electric field created by any group of charges is everywhere perpendicular to the associated equipotential surfaces and points in the direction of decreasing potential.

The electric field is related to two equipotential surfaces by

$$E = -\frac{\Delta V}{\Delta s} \qquad (19.7a)$$

where ΔV is the potential difference between the surfaces and Δs is the displacement. The term $\Delta V/\Delta s$ is called the potential gradient.

19.5 | Capacitors and Dielectrics

A capacitor is a device that stores charge and energy. It consists of two conductors or plates that are near one another, but not touching. The magnitude q of the charge on each plate is given by

$$q = CV \qquad (19.8)$$

where V is the magnitude of the potential difference between the plates and C is the capacitance. The SI unit for capacitance is the coulomb per volt (C/V), or farad (F).

The insulating material included between the plates of a capacitor is called a dielectric. The dielectric constant κ of the material is defined as

$$\kappa = \frac{E_0}{E} \qquad (19.9)$$

where E_0 and E are, respectively, the magnitudes of the electric fields between the plates without and with a dielectric, assuming the charge on the plates is kept fixed.

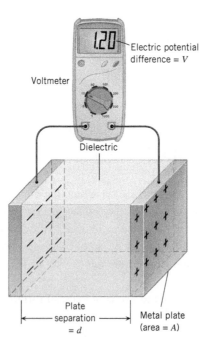

| FIGURE 19.17

A parallel plate capacitor consists of two metal plates, one carrying a charge $+q$ and the other a charge $-q$. The potential of the positive plate exceeds that of the negative plate by an amount V. The region between the plates is filled with a dielectric.

The capacitance of a parallel plate capacitor filled with a dielectric is

$$C = \frac{\kappa \epsilon_0 A}{d} \tag{19.10}$$

where $\epsilon_0 = 8.85 \times 10^{-12}\,\text{C}^2/(\text{N} \cdot \text{m}^2)$ is the permittivity of free space, A is the area of each plate, and d is the distance between the plates.

The electric potential energy stored in a capacitor is

$$\text{Energy} = \tfrac{1}{2}qV = \tfrac{1}{2}CV^2 = q^2/(2C) \tag{19.11a–c}$$

The energy density is the energy stored per unit volume and is related to the magnitude E of the electric field as follows:

$$\text{Energy density} = \tfrac{1}{2}\kappa \epsilon_0 E^2 \tag{19.12}$$

CHAPTER 20

Electric Circuits

20.1 | Electromotive Force and Current

There must be at least one source or generator of electrical energy in an electric circuit. The electromotive force (emf) of a generator, such as a battery, is the maximum potential difference (in volts) that exists between the terminals of the generator.

The rate of flow of charge is called the electric current. If the rate is constant, the current I is given by

$$I = \frac{\Delta q}{\Delta t} \qquad \text{(20.1)}$$

where Δq is the magnitude of the charge crossing a surface in a time Δt, the surface being perpendicular to the motion of the charge. The SI unit for current is the coulomb per second (C/s), which is referred to as an ampere (A).

When the charges flow only in one direction around a circuit, the current is called direct current (dc). When the direction of charge flow changes from moment to moment, the current is known as alternating current (ac).

Conventional current is the hypothetical flow of positive charges that would have the same effect in a circuit as the movement of negative charges that actually does occur.

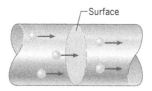

| FIGURE 20.3

The electric current is the amount of charge per unit time that passes through an imaginary surface that is perpendicular to the motion of the charges.

20.2 | Ohm's Law

The definition of electrical resistance R is $R = V/I$, where V (in volts) is the voltage applied across a piece of material and I (in amperes) is the current through the material. Resistance is measured in volts per ampere, a unit called an ohm (Ω).

If the ratio of the voltage to the current is constant for all values of voltage and current, the resistance is constant. In this event, the definition of resistance becomes Ohm's law, as follows:

$$\frac{V}{I} = R = \text{constant} \quad \text{or} \quad V = IR \tag{20.2}$$

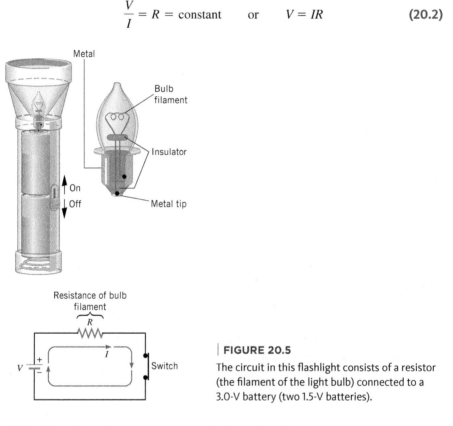

| FIGURE 20.5
The circuit in this flashlight consists of a resistor (the filament of the light bulb) connected to a 3.0-V battery (two 1.5-V batteries).

20.3 | Resistance and Resistivity

The resistance of a piece of material of length L and cross-sectional area A is

$$R = \rho \frac{L}{A} \tag{20.3}$$

where ρ is the resistivity of the material.

The resistivity of a material depends on the temperature. For many materials and limited temperature ranges, the temperature dependence is given by

$$\rho = \rho_0[1 + \alpha(T - T_0)] \qquad (20.4)$$

where ρ and ρ_0 are the resistivities at temperatures T and T_0, respectively, and α is the temperature coefficient of resistivity.

The temperature dependence of the resistance R is given by

$$R = R_0[1 + \alpha(T - T_0)] \qquad (20.5)$$

where R and R_0 are the resistances at temperatures T and T_0, respectively.

TABLE 20-1

Resistivities[a] of Various Materials

Material	Resistivity ρ ($\Omega \cdot m$)	Material	Resistivity ρ ($\Omega \cdot m$)
Conductors		*Semiconductors*	
Aluminum	2.82×10^{-8}	Carbon	3.5×10^{-5}
Copper	1.72×10^{-8}	Germanium	0.5^{b}
Gold	2.44×10^{-8}	Silicon	$20 - 2300^{b}$
Iron	9.7×10^{-8}	*Insulators*	
Mercury	95.8×10^{-8}	Mica	10^{11}–10^{15}
Nichrome (alloy)	100×10^{-8}	Rubber (hard)	10^{13}–10^{16}
Silver	1.59×10^{-8}	Teflon	10^{16}
Tungsten	5.6×10^{-8}	Wood (maple)	3×10^{10}

[a]The values pertain to temperatures near 20 °C.
[b]Depending on purity.

20.4 | Electric Power

When electric charge flows from point A to point B in a circuit, leading to a current I, and the voltage between the points is V, the electric power associated with this current and voltage is

$$P = IV \qquad (20.6a)$$

For a resistor, Ohm's law applies, and it follows that the power delivered to the resistor is also given by either of the following two equations:

$$P = I^2R \qquad (20.6b)$$

$$P = \frac{V^2}{R} \qquad (20.6c)$$

20.5 | Alternating Current

The alternating voltage between the terminals of an ac generator can be represented by

$$V = V_0 \sin 2\pi ft \qquad (20.7)$$

where V_0 is the peak value of the voltage, t is the time, and f is the frequency (in Hertz) at which the voltage oscillates.

Correspondingly, in a circuit containing only resistance, the ac current is

$$I = I_0 \sin 2\pi ft \qquad (20.8)$$

where I_0 is the peak value of the current and is related to the peak voltage via $I_0 = V_0/R$.

For sinusoidal current and voltage, the root mean square (rms) current and voltage are related to the peak values according to the following equations:

$$I_{rms} = \frac{I_0}{\sqrt{2}} \qquad (20.12)$$

$$V_{rms} = \frac{V_0}{\sqrt{2}} \qquad (20.13)$$

The power in an ac circuit is the product of the current and the voltage and oscillates in time. The average power is

$$\overline{P} = I_{rms}V_{rms} \qquad (20.15a)$$

For a resistor, Ohm's law applies, so that $V_{rms} = I_{rms}R$ and the average power delivered to the resistor is also given by the following two equations:

$$\overline{P} = I_{rms}^2 R \qquad (20.15b)$$

$$\overline{P} = \frac{V_{rms}^2}{R} \qquad (20.15c)$$

20.6 | Series Wiring

When devices are connected in series, there is the same current through each device. The equivalent resistance R_S of a series combination of resistances (R_1, R_2, R_3, etc.) is

$$R_S = R_1 + R_2 + R_3 + \cdots \qquad (20.16)$$

The power delivered to the equivalent resistance is equal to the total power delivered to any number of resistors in series.

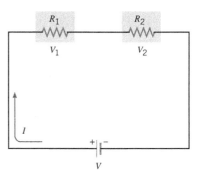

| FIGURE 20.14

When two resistors are connected in series, the same current I is in both of them.

20.7 | Parallel Wiring

When devices are connected in parallel, the same voltage is applied across each device. In general, devices wired in parallel carry different currents. The reciprocal of the equivalent resistance R_P of a parallel combination of resistances (R_1, R_2, R_3, etc.) is

$$\frac{1}{R_P} = \frac{1}{R_1} + \frac{1}{R_2} + \frac{1}{R_3} + \cdots \tag{20.17}$$

The power delivered to the equivalent resistance is equal to the total power delivered to any number of resistors in parallel.

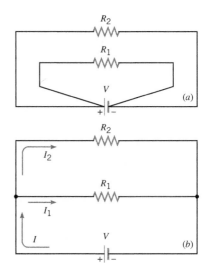

| FIGURE 20.18

(a) When two resistors are connected in parallel, the same voltage V is applied across each resistor.

(b) This drawing is equivalent to part a. I_1 and I_2 are the currents in R_1 and R_2.

20.8 | Circuits Wired Partially in Series and Partially in Parallel

Sometimes, one section of a circuit is wired in series, while another is wired in parallel. In such cases the circuit can be analyzed in parts, according to the respective series and parallel equivalent resistances of the various sections.

20.9 | Internal Resistance

The internal resistance of a battery or generator is the resistance within the battery or generator. The terminal voltage is the voltage between the terminals of a battery or generator and is equal to the emf only when there is no current through the device. When there is a current I, the internal resistance r causes the terminal voltage to be less than the emf by an amount Ir.

20.10 | Kirchhoff's Rules

Kirchhoff's junction rule states that the sum of the magnitudes of the currents directed into a junction equals the sum of the magnitudes of the currents directed out of the junction.

Kirchhoff's loop rule states that, around any closed-circuit loop, the sum of the potential drops equals the sum of the potential rises.

REASONING STRATEGY
Applying Kirchhoff's Rules

1 Draw the current in each branch of the circuit. Choose any direction. If your choice is incorrect, the value obtained for the current will turn out to be a negative number.

2 Mark each resistor with a plus sign at one end and a minus sign at the other end, in a way that is consistent with your choice for the current direction in Step 1. Outside a battery, conventional current is always directed from a higher potential (the end marked +) toward a lower potential (the end marked −).

3 Apply the junction rule and the loop rule to the circuit, obtaining in the process as many independent equations as there are unknown variables.

4 Solve the equations obtained in Step 3 simultaneously for the unknown variables. (See Chapter 0, section 0.9.)

20.11 | The Measurement of Current and Voltage

A galvanometer is a device that responds to electric current and is used in nondigital ammeters and voltmeters. An ammeter is an instrument that measures current and must be inserted into a circuit in such a way that the current passes directly through the ammeter. A voltmeter is an instrument for measuring the voltage between two points in a circuit. A voltmeter must be connected between the two points and is not inserted into a circuit as an ammeter is.

20.12 | Capacitors in Series and in Parallel

The equivalent capacitance C_P for a parallel combination of capacitances (C_1, C_2, C_3, etc.) is

$$C_P = C_1 + C_2 + C_3 + \cdots \qquad (20.18)$$

In general, each capacitor in a parallel combination carries a different amount of charge. The equivalent capacitor carries the same total charge and stores the same total energy as the parallel combination.

The reciprocal of the equivalent capacitance C_S for a series combination (C_1, C_2, C_3, etc.) of capacitances is

$$\frac{1}{C_S} = \frac{1}{C_1} + \frac{1}{C_2} + \frac{1}{C_3} + \cdots \qquad (20.19)$$

The equivalent capacitor carries the same amount of charge as *any one* of the capacitors in the combination and stores the same total energy as the series combination.

20.13 | RC Circuits

The charging or discharging of a capacitor in a dc series circuit (resistance R, capacitance C) does not occur instantaneously. The charge on a capacitor builds up gradually, as described by the following equation:

$$q = q_0[1 - e^{-t/(RC)}] \qquad (20.20)$$

where q is the charge on the capacitor at time t and q_0 is the equilibrium value of the charge. The time constant τ of the circuit is

$$\tau = RC \qquad (20.21)$$

The discharging of a capacitor through a resistor is described as follows:

$$q = q_0 e^{-t/(RC)} \qquad (20.22)$$

where q_0 is the charge on the capacitor at time $t = 0$ s.

CHAPTER 21

Magnetic Forces and Magnetic Fields

21.1 Magnetic Fields

A magnet has a north pole and a south pole. The north pole is the end that points toward the north magnetic pole of the earth when the magnet is freely suspended. Like magnetic poles repel each other, and unlike poles attract each other.

A magnetic field exists in the space around a magnet. The magnetic field is a vector whose direction at any point is the direction indicated by the north pole of a small compass needle placed at that point.

As an aid in visualizing the magnetic field, magnetic field lines are drawn in the vicinity of a magnet. The lines appear to originate from the north pole and end on the south pole. The magnetic field at any point in space is tangent to the magnetic field line at the point. Furthermore, the strength of the magnetic field is proportional to the number of lines per unit area that passes through a surface oriented perpendicular to the lines.

See Figure 21.7 on the next page.

21.2 | The Force That a Magnetic Field Exerts on a Moving Charge

The direction of the magnetic force acting on a charge moving with a velocity $\vec{v}$ in a magnetic field $\vec{B}$ is perpendicular to both $\vec{v}$ and $\vec{B}$. For a positive charge the

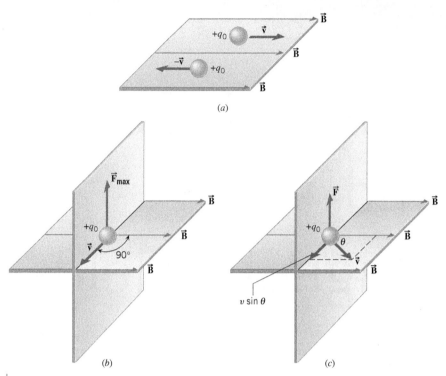

(a)

(b)

(c)

| **FIGURE 21.7**

(a) No magnetic force acts on a charge moving with a velocity $\vec{v}$ that is parallel or antiparallel to a magnetic field $\vec{B}$.

(b) The charge experiences a maximum force $\vec{F}_{max}$ when the charge moves perpendicular to the field.

(c) If the charge travels at an angle θ with respect to $\vec{B}$, only the velocity component perpendicular to $\vec{B}$ gives rise to a magnetic force $\vec{F}$, which is smaller than $\vec{F}_{max}$. This component is $v \sin \theta$.

direction can be determined with the aid of Right-Hand Rule No. 1 (see Figure 21.8). The magnetic force on a moving negative charge is opposite to the force on a moving positive charge.

Right-Hand Rule No. 1 Extend the right hand so the fingers point along the direction of the magnetic field $\vec{B}$ and the thumb points along the velocity $\vec{v}$ of the charge. The palm of the hand then faces in the direction of the magnetic force $\vec{F}$ that acts on a positive charge.

The magnitude B of the magnetic field at any point in space is defined as

$$B = \frac{F}{|q_0| v \sin \theta}$$

(21.1)

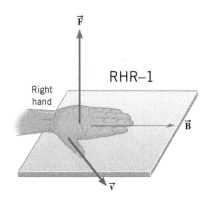

RHR–1

| **FIGURE 21.8**

Right-Hand Rule No. 1 (RHR–1) is illustrated. When the right hand is oriented so the fingers point along the magnetic field $\vec{B}$ and the thumb points along the velocity $\vec{v}$ of a positively charged particle, the palm faces in the direction of the magnetic force $\vec{F}$ applied to the particle.

where F is the magnitude of the magnetic force on a test charge, $|q_0|$ is the magnitude of the test charge, and v is the magnitude of the charge's velocity, which makes an angle θ with the direction of the magnetic field. The SI unit for the magnetic field is the tesla (T). Another, smaller unit for the magnetic field is the gauss; 1 gauss = 10^{-4} tesla. The gauss is not an SI unit.

21.3 The Motion of a Charged Particle in a Magnetic Field

When a charged particle moves in a region that contains both magnetic and electric fields, the net force on the particle is the vector sum of the magnetic and electric forces.

A magnetic force does no work on a particle, because the direction of the force is always perpendicular to the motion of the particle. Being unable to do work, the magnetic force cannot change the kinetic energy, and hence the speed, of the particle; however, the magnetic force does change the direction in which the particle moves.

When a particle of charge q (magnitude = $|q|$) and mass m moves with speed v perpendicular to a uniform magnetic field of magnitude B, the magnetic force causes the particle to move on a circular path of radius

$$r = \frac{mv}{|q_0|B} \tag{21.2}$$

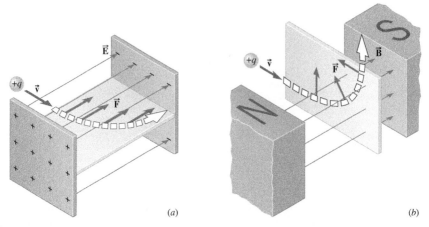

| FIGURE 21.9
(a) The electric force $\vec{F}$ that acts on a positive charge is parallel to the electric field $\vec{E}$.
(b) The magnetic force $\vec{F}$ is perpendicular to both the magnetic field $\vec{B}$ and the velocity $\vec{v}$.

21.4 | The Mass Spectrometer

The mass spectrometer is an instrument for measuring the abundance of ionized atoms or molecules that have different masses. The atoms or molecules are ionized $(+e)$, accelerated to a speed v by a potential difference V, and sent into a uniform magnetic field of magnitude B. The magnetic field causes the particles (each with a mass m) to move on a circular path of radius r. The relation between m and B is

$$m = \left(\frac{er^2}{2V}\right) B^2$$

21.5 | The Force on a Current in a Magnetic Field

An electric current, being composed of moving charges, can experience a magnetic force when placed in a magnetic field of magnitude B. For a straight wire that has a length L and carries a current I, the magnetic force has a magnitude of

$$F = ILB \sin \theta \qquad (21.3)$$

where θ is the angle between the directions of the current and the magnetic field. The direction of the force is perpendicular to both the current and the magnetic field and is given by Right-Hand Rule No. 1.

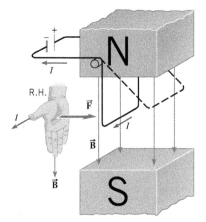

| **FIGURE 21.16**

The wire carries a current I, and the bottom segment of the wire is oriented perpendicular to a magnetic field $\vec{B}$. A magnetic force deflects the wire to the right.

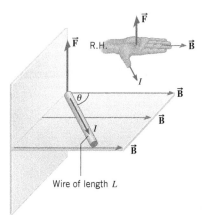

| **FIGURE 21.17**

The current I in the wire, oriented at an angle θ with respect to a magnetic field $\vec{B}$, is acted upon by a magnetic force $\vec{F}$.

21.6 | The Torque on a Current-Carrying Coil

Magnetic forces can exert a torque on a current-carrying loop of wire and thus cause the loop to rotate. When a current I exists in a coil of wire with N turns, each of area A, in the presence of a magnetic field of magnitude B, the coil experiences a net torque of magnitude

$$\tau = NIAB \sin \phi \qquad\qquad (21.4)$$

where ϕ is the angle between the direction of the magnetic field and the normal to the plane of the coil.

The quantity NIA is known as the magnetic moment of the coil.

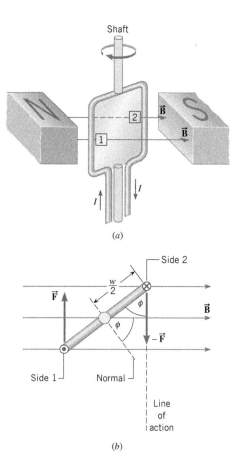

(a)

(b)

| FIGURE 21.19

(a) A current-carrying loop of wire, which can rotate about a vertical shaft, is situated in a magnetic field.

(b) A top view of the loop. The current in side 1 is directed out of the page (⊙), while the current in side 2 is directed into the page (⊗). The current in side 1 experiences a force $\vec{F}$ that is opposite to the force exerted on side 2. The two forces produce a clockwise torque about the shaft.

21.7 | Magnetic Fields Produced by Currents

An electric current produces a magnetic field, with different current geometries giving rise to different field patterns. For an infinitely long, straight wire, the magnetic field lines are circles centered on the wire, and their direction is given by Right-Hand Rule No. 2 (see below). The magnitude of the magnetic field at a radial distance r from the wire is

$$B = \frac{\mu_0 I}{2\pi r} \tag{21.5}$$

where I is the current in the wire and μ_0 is a constant known as the permeability of free space ($\mu_0 = 4\pi \times 10^{-7}\,\text{T}\cdot\text{m/A}$).

Right-Hand Rule No. 2 Curl the fingers of the right hand into the shape of a half-circle. Point the thumb in the direction of the conventional current I, and the tips of the fingers will point in the direction of the magnetic field $\vec{\mathbf{B}}$.

The magnitude of the magnetic field at the center of a flat circular loop consisting of N turns, each of radius R, is

$$B = N\frac{\mu_0 I}{2R} \tag{21.6}$$

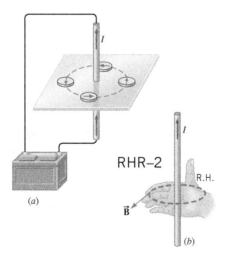

| **FIGURE 21.23**
(a) A very long, straight, current-carrying wire produces magnetic field lines that are circular about the wire, as indicated by the compass needles.
(b) With the thumb of the right hand (R.H.) along the current I, the curled fingers point in the direction of the magnetic field, according to Right-Hand Rule No. 2 (RHR-2).

The loop has associated with it a north pole on one side and a south pole on the other side. The side of the loop that behaves like a north pole can be predicted by using Right-Hand Rule No. 2.

A solenoid is a coil of wire wound in the shape of a helix. Inside a long solenoid the magnetic field is nearly constant and has a magnitude of

$$B = \mu_0 n I \tag{21.7}$$

where n is the number of turns per unit length of the solenoid. One end of the solenoid behaves like a north pole, and the other end like a south pole. The end that is the north pole can be predicted by using Right-Hand Rule No. 2.

21.8 | Ampère's Law

Ampère's law specifies the relationship between a current and its associated magnetic field. For any current geometry that produces a magnetic field that does not change in time, Ampère's law states that

$$\Sigma B_\parallel \Delta \ell = \mu_0 I \tag{21.8}$$

where $\Delta \ell$ is a small segment of length along a closed path of arbitrary shape around the current, $B_\parallel$ is the component of the magnetic field parallel to $\Delta \ell$, I is the net current passing through the surface bounded by the path, and μ_0 is the permeability of free space. The symbol Σ indicates that the sum of all $B_\parallel \Delta \ell$ terms must be taken around the closed path.

21.9 | Magnetic Materials

Ferromagnetic materials, such as iron, are made up of tiny regions called magnetic domains, each of which behaves as a small magnet. In an unmagnetized ferromagnetic material, the domains are randomly aligned. In a permanent magnet, many of the domains are aligned, and a high degree of magnetism results. An unmagnetized ferromagnetic material can be induced into becoming magnetized by placing it in an external magnetic field.

CHAPTER 22

Electromagnetic Induction

22.1 | Induced Emf and Induced Current

Electromagnetic induction is the phenomenon in which an emf is induced in a piece of wire or a coil of wire with the aid of a magnetic field. The emf is called an induced emf, and any current that results from the emf is called an induced current.

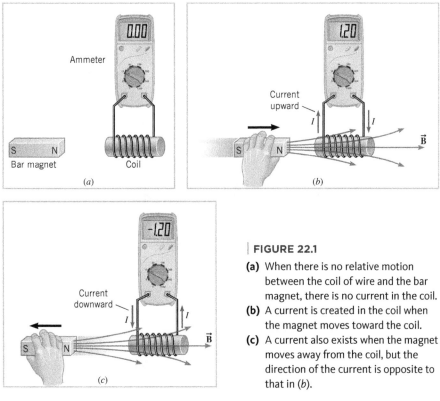

FIGURE 22.1

(a) When there is no relative motion between the coil of wire and the bar magnet, there is no current in the coil.

(b) A current is created in the coil when the magnet moves toward the coil.

(c) A current also exists when the magnet moves away from the coil, but the direction of the current is opposite to that in (b).

22.2 | Motional Emf

An emf $\mathcal{E}$ is induced in a conducting rod of length L when the rod moves with a speed v in a magnetic field of magnitude B, according to

$$\mathcal{E} = vBL \qquad\qquad\qquad (22.1)$$

Equation 22.1 applies when the velocity of the rod, the length of the rod, and the magnetic field are mutually perpendicular.

When the motional emf is used to operate an electrical device, such as a light bulb, the energy delivered to the device originates in the work done to move the rod, and the law of conservation of energy applies.

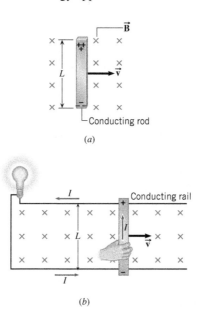

(a)

(b)

| FIGURE 22.4

(a) When a conducting rod moves at right angles to a constant magnetic field, the magnetic force causes opposite charges to appear at the ends of the rod, giving rise to an induced emf.

(b) The induced emf causes an induced current I to appear in the circuit.

22.3 | Magnetic Flux

The magnetic flux Φ that passes through a surface is

$$\Phi = BA \cos \phi \qquad\qquad\qquad (22.2)$$

where B is the magnitude of the magnetic field, A is the area of the surface, and ϕ is the angle between the field and the normal to the surface.

The magnetic flux is proportional to the number of magnetic field lines that pass through the surface.

22.4 | Faraday's Law of Electromagnetic Induction

Faraday's law of electromagnetic induction states that the average emf $\mathscr{E}$ induced in a coil of N loops is

$$\mathscr{E} = -N\left(\frac{\Phi - \Phi_0}{t - t_0}\right) = -N\frac{\Delta\Phi}{\Delta t} \qquad (22.3)$$

where $\Delta\Phi$ is the change in magnetic flux through one loop and Δt is the time interval during which the change occurs. Motional emf is a special case of induced emf.

22.5 | Lenz's Law

Lenz's law provides a way to determine the polarity of an induced emf. Lenz's law is stated as follows: The induced emf resulting from a changing magnetic flux has a polarity that leads to an induced current whose direction is such that the induced magnetic field opposes the original flux change. This statement is a consequence of the law of conservation of energy.

✴❓ REASONING STRATEGY
Determining the Polarity of the Induced Emf

1. Determine whether the magnetic flux that penetrates a coil is increasing or decreasing.
2. Find what the direction of the induced magnetic field must be so that it can *oppose the change in flux* by adding to or subtracting from the original field.
3. Having found the direction of the induced magnetic field, use RHR-2 (see Section 21.7) to determine the direction of the induced current. Then the polarity of the induced emf can be assigned because conventional current is directed out of the positive terminal, through the external circuit, and into the negative terminal.

22.7 | The Electric Generator

In its simplest form, an electric generator consists of a coil of N loops that rotates in a uniform magnetic field $\vec{B}$. The emf produced by this generator is

$$\mathcal{E} = NAB\omega \sin \omega t = \mathcal{E}_0 \sin \omega t \qquad (22.4)$$

where A is the area of the coil, ω is the angular speed (in rad/s) of the coil, and $\mathcal{E}_0 = NAB\omega$ is the peak emf. The angular speed in rad/s is related to the frequency f in cycles/s, or Hz, according to $\omega = 2\pi f$.

When an electric motor is running, it exhibits a generator-like behavior by producing an induced emf, called the back emf. The current I needed to keep the motor running at a constant speed is

$$I = \frac{V - \mathcal{E}}{R} \qquad (22.5)$$

where V is the emf applied to the motor by an external source, $\mathcal{E}$ is the back emf, and R is the resistance of the motor coil.

22.8 | Mutual Inductance and Self-Inductance

Mutual induction is the effect in which a changing current in the primary coil induces an emf in the secondary coil. The average emf $\mathcal{E}_s$ induced in the secondary coil by a change in current ΔI_p in the primary coil is

$$\mathcal{E}_s = -M \frac{\Delta I_p}{\Delta t} \qquad (22.7)$$

where Δt is the time interval during which the change occurs. The constant M is the mutual inductance between the two coils and is measured in henries (H).

Self-induction is the effect in which a change in current ΔI in a coil induces an average emf $\mathcal{E}$ in the same coil, according to

$$\mathcal{E} = -L \frac{\Delta I}{\Delta t} \qquad (22.9)$$

The constant L is the self-inductance, or inductance, of the coil and is measured in henries.

To establish a current I in an inductor, work must be done by an external agent. This work is stored as energy in the inductor, the amount of energy being

$$\text{Energy} = \tfrac{1}{2}LI^2 \tag{22.10}$$

The energy stored in an inductor can be regarded as being stored in its magnetic field. At any point in air or vacuum or in a nonmagnetic material where a magnetic field $\vec{B}$ exists, the energy density, or the energy stored per unit volume, is

$$\text{Energy density} = \frac{1}{2\mu_0}B^2 \tag{22.11}$$

22.9 | Transformers

A transformer consists of a primary coil of N_p turns and a secondary coil of N_s turns. If the resistances of the coils are negligible, the voltage V_p across the primary coil and the voltage V_s across the secondary coil are related according to the transformer equation:

$$\frac{V_s}{V_p} = \frac{N_s}{N_p} \tag{22.12}$$

where the ratio N_s/N_p is called the turns ratio of the transformer.

A transformer functions with ac electricity, not with dc. If the transformer is 100% efficient in transferring power from the primary to the secondary coil, the ratio of the secondary current I_s to the primary current I_p is

$$\frac{I_s}{I_p} = \frac{N_p}{N_s} \tag{22.13}$$

CHAPTER 23

Alternating Current Circuits

23.1 | Capacitors and Capacitive Reactance

In an ac circuit the rms voltage V_{rms} across a capacitor is related to the rms current I_{rms} by

$$V_{rms} = I_{rms}X_C \tag{23.1}$$

where X_C is the capacitive reactance. The capacitive reactance is measured in ohms (Ω) and is given by

$$X_C = \frac{1}{2\pi f C} \tag{23.2}$$

where f is the frequency and C is the capacitance.

    The ac current in a capacitor leads the voltage across the capacitor by a phase angle of $90°$ or $\pi/2$ radians. As a result, a capacitor consumes no power, on average.

    The phasor model is useful for analyzing the voltage and current in an ac circuit. In this model, the voltage and current are represented by rotating arrows, called phasors.

    The length of the voltage phasor represents the maximum voltage V_0, and the length of the current phasor represents the maximum current I_0. The phasors rotate in a counterclockwise direction at a frequency f. Since the current leads the voltage by $90°$ in a capacitor, the current phasor is ahead of the voltage phasor by $90°$ in the direction of rotation.

    The instantaneous values of the voltage and current are equal to the vertical components of the corresponding phasors.

    See Figure 23.5 on the next page.

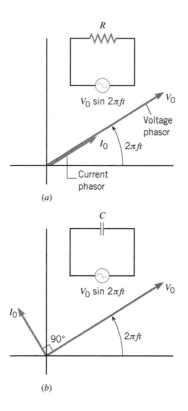

| FIGURE 23.5

These rotating-arrow models represent the voltage and the current in ac circuits that contain (a) only a resistor and (b) only a capacitor.

23.2 | Inductors and Inductive Reactance

In an ac circuit the rms voltage V_{rms} across an inductor is related to the rms current I_{rms} by

$$V_{rms} = I_{rms}X_L \tag{23.3}$$

where X_L is the inductive reactance. The inductive reactance is measured in ohms (Ω) and is given by

$$X_L = 2\pi f L \tag{23.4}$$

where f is the frequency and L is the inductance.

The ac current in an inductor lags behind the voltage across the inductor by a phase angle of 90° or $\pi/2$ radians. Consequently, an inductor, like a capacitor, consumes no power, on average.

The voltage and current phasors in a circuit containing only an inductor also rotate in a counterclockwise direction at a frequency f. However, since the current lags the voltage by 90° in an inductor, the current phasor is behind the voltage phasor by 90° in the direction of rotation.

The instantaneous values of the voltage and current are equal to the vertical components of the corresponding phasors.

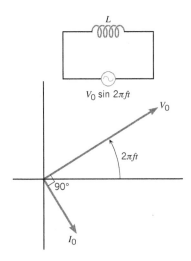

| **FIGURE 23.8**

This phasor model represents the voltage and current in a circuit that contains only an inductor.

23.3 | Circuits Containing Resistance, Capacitance, and Inductance

When a resistor, a capacitor, and an inductor are connected in series, the rms voltage across the combination is related to the rms current according to

$$V_{rms} = I_{rms}Z \tag{23.6}$$

where Z is the impedance of the combination. The impedance is measured in ohms (Ω) and is given by

$$Z = \sqrt{R^2 + (X_L - X_C)^2} \tag{23.7}$$

where R is the resistance, and X_L and X_C are, respectively, the inductive and capacitive reactances.

The tangent of the phase angle ϕ between current and voltage in a series RCL circuit is

$$\tan \phi = \frac{X_L - X_C}{R} \tag{23.8}$$

Only the resistor in the RCL combination consumes power, on average. The average power $\overline{P}$ consumed in the circuit is

$$\overline{P} = I_{rms} V_{rms} \cos \phi \tag{23.9}$$

where $\cos \phi$ is called the power factor of the circuit.

23.4 | Resonance in Electric Circuits

A series RCL circuit has a resonant frequency f_0 that is given by

$$f_0 = \frac{1}{2\pi\sqrt{LC}} \tag{23.10}$$

where L is the inductance and C is the capacitance. At resonance, the impedance of the circuit has a minimum value equal to the resistance R, and the rms current has a maximum value.

23.5 | Semiconductor Devices

In an n-type semiconductor, mobile negative electrons carry the current. An n-type material is produced by doping a semiconductor such as silicon with a small amount of impurity atoms such as phosphorus.

In a p-type semiconductor, mobile positive "holes" in the crystal structure carry the current. A p-type material is produced by doping a semiconductor with a small amount of impurity atoms such as boron.

These two types of semiconductors are used in p-n junction diodes, light-emitting diodes, and solar cells, and in pnp and npn bipolar junction transistors.

CHAPTER 24

Electromagnetic Waves

24.1 | The Nature of Electromagnetic Waves

An electromagnetic wave consists of mutually perpendicular and oscillating electric and magnetic fields. The wave is a transverse wave, since the fields are perpendicular to the direction in which the wave travels. Electromagnetic waves can travel through a vacuum or a material substance. All electromagnetic waves travel through a vacuum at the same speed, which is known as the speed of light c ($c = 3.00 \times 10^8$ m/s).

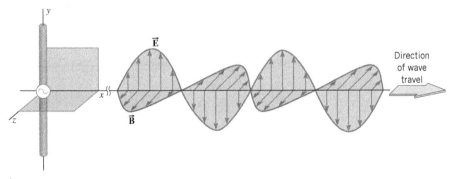

| FIGURE 24.3

This picture shows the wave of the radiation field far from the antenna. Observe that $\vec{E}$ and $\vec{B}$ are perpendicular to each other, and both are perpendicular to the direction of travel.

24.2 | The Electromagnetic Spectrum

The frequency f and wavelength λ of an electromagnetic wave in a vacuum are related to its speed c through the relation

$$c = f\lambda$$

The series of electromagnetic waves, arranged in order of their frequencies or wavelengths, is called the electromagnetic spectrum. In increasing order of frequency (decreasing order of wavelength), the spectrum includes radio waves, infrared radiation, visible light, ultraviolet radiation, X-rays, and gamma rays. Visible light has frequencies between about 4.0×10^{14} and 7.9×10^{14} Hz. The human eye and brain perceive different frequencies or wavelengths as different colors.

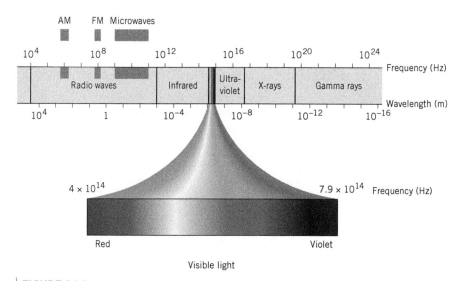

| FIGURE 24.9
The electromagnetic spectrum.

24.3 | The Speed of Light

James Clerk Maxwell showed that the speed of light in a vacuum is

$$c = \frac{1}{\sqrt{\epsilon_0 \mu_0}} \tag{24.1}$$

where ϵ_0 is the (electric) permittivity of free space and μ_0 is the (magnetic) permeability of free space.

24.4 | The Energy Carried by Electromagnetic Waves

The total energy density u of an electromagnetic wave is the total energy per unit volume of the wave and, in a vacuum, is given by

$$u = \frac{1}{2}\epsilon_0 E^2 + \frac{1}{2\mu_0}B^2 \qquad (24.2a)$$

where E and B, respectively, are the magnitudes of the electric and magnetic fields of the wave. Since the electric and magnetic parts of the total energy density are equal, the following two equations are equivalent to Equation 24.2a:

$$u = \epsilon_0 E^2 \qquad (24.2b)$$

$$u = \frac{1}{\mu_0}B^2 \qquad (24.2c)$$

In a vacuum, E and B are related according to

$$E = cB \qquad (24.3)$$

Equations 24.2a, 24.2b, and 24.2c can be used to determine the average total energy density, if the rms average values E_{rms} and B_{rms} are used in place of the symbols E and B. The rms values are related to the peak values E_0 and B_0 in the usual way:

$$E_{rms} = \frac{1}{\sqrt{2}}E_0 \quad \text{and} \quad B_{rms} = \frac{1}{\sqrt{2}}B_0$$

The intensity of an electromagnetic wave is the power that the wave carries perpendicularly through a surface divided by the area of the surface. In a vacuum, the intensity S is related to the total energy density u according to

$$S = cu \qquad (24.4)$$

24.5 | The Doppler Effect and Electromagnetic Waves

When electromagnetic waves and the source and observer of the waves all travel along the same line in a vacuum, the Doppler effect is given by

$$f_o = f_s\left(1 \pm \frac{v_{rel}}{c}\right) \quad \text{if } v_{rel} \ll c \qquad (24.6)$$

where f_o and f_s are, respectively, the observed and emitted wave frequencies and v_{rel} is the relative speed of the source and the observer. The plus sign is used when the source and the observer come together, and the minus sign is used when they move apart.

24.6 | Polarization

A linearly polarized electromagnetic wave is one in which the oscillation of the electric field occurs only along one direction, which is taken to be the direction of polarization. The magnetic field also oscillates along only one direction, which is perpendicular to the electric field direction. In an unpolarized wave such as the light from an incandescent bulb, the direction of polarization does not remain fixed, but fluctuates randomly in time.

Polarizing materials allow only the component of the wave's electric field along one direction (and the associated magnetic field component) to pass through them. The preferred transmission direction for the electric field is called the transmission axis of the material.

When unpolarized light is incident on a piece of polarizing material, the transmitted polarized light has an average intensity that is one-half the average intensity of the incident light.

When two pieces of polarizing material are used one after the other, the first is called the polarizer, and the second is referred to as the analyzer. If the average intensity of polarized light falling on the analyzer is $\overline{S}_0$, the average intensity $\overline{S}$ of the light leaving the analyzer is given by Malus' law as

$$\overline{S} = \overline{S}_0 \cos^2 \theta \qquad (24.7)$$

where θ is the angle between the transmission axes of the polarizer and analyzer. When $\theta = 90°$, the polarizer and the analyzer are said to be "crossed," and no light passes through the analyzer.

CHAPTER 25

The Reflection of Light: Mirrors

25.1 | Wave Fronts and Rays

Wave fronts are surfaces on which all points of a wave are in the same phase of motion. Waves whose wave fronts are flat surfaces are known as plane waves.

Rays are lines that are perpendicular to the wave fronts and point in the direction of the velocity of the wave.

25.2 | The Reflection of Light

When light reflects from a smooth surface, the reflected light obeys the law of reflection:

(a) The incident ray, the reflected ray, and the normal to the surface all lie in the same plane.

(b) The angle of reflection θ_r equals the angle of incidence θ_i; $\theta_r = \theta_i$.

25.3 | The Formation of Images by a Plane Mirror

A virtual image is one from which all the rays of light do not actually come, but only appear to do so.

A real image is one from which all the rays of light actually do emanate.

A plane mirror forms an upright, virtual image that is located as far behind the mirror as the object is in front of it. In addition, the heights of the image and the object are equal.

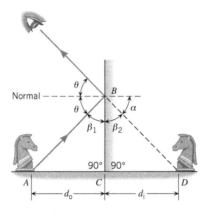

| FIGURE 25.7

This drawing illustrates the geometry used with a plane mirror to show that the image distance d_i equals the object distance d_o.

25.4 | Spherical Mirrors

A spherical mirror has the shape of a section from the surface of a sphere. If the inside surface of the mirror is polished, it is a concave mirror. If the outside surface is polished, it is a convex mirror.

The principal axis of a mirror is a straight line drawn through the center of curvature and the middle of the mirror's surface. Rays that are close to the principal axis are known as paraxial rays. Paraxial rays are not necessarily parallel to the principal axis.

The radius of curvature R of a mirror is the distance from the center of curvature to the mirror.

The focal point of a concave spherical mirror is a point on the principal axis, in front of the mirror. Incident paraxial rays that are parallel to the principal axis converge at the focal point after being reflected from the concave mirror.

The focal point of a convex spherical mirror is a point on the principal axis, behind the mirror. For a convex mirror, incident paraxial rays that are parallel to the principal axis diverge after reflecting from the mirror. These rays seem to originate from the focal point.

The fact that a spherical mirror does not bring all rays parallel to the principal axis to a single image point after reflection is known as spherical aberration.

The focal length f of a mirror is the distance along the principal axis between the focal point and the mirror. The focal length and the radius of curvature R are related by

$$f = \tfrac{1}{2}R \tag{25.1}$$

$$f = -\tfrac{1}{2}R \tag{25.2}$$

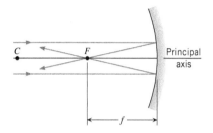

| FIGURE 25.12

Light rays near and parallel to the principal axis are reflected from a concave mirror and converge at the focal point F. The focal length f is the distance between F and the mirror.

25.5 | The Formation of Images by Spherical Mirrors

The image produced by a mirror can be located by a graphical method known as ray tracing.

For a concave mirror, the following paraxial rays are especially useful for ray tracing (see Figure 25.17 on the next page):

Ray 1. This ray leaves the object traveling parallel to the principal axis. The ray reflects from the mirror and passes through the focal point.

Ray 2. This ray leaves the object and passes through the focal point. The ray reflects from the mirror and travels parallel to the principal axis.

Ray 3. This ray leaves the object and travels along a line that passes through the center of curvature. The ray strikes the mirror perpendicularly and reflects back on itself.

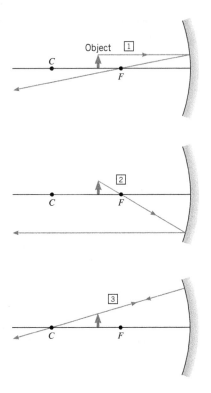

| **FIGURE 25.17**
The rays labeled 1, 2, and 3 are useful in locating the image of an object placed in front of a concave spherical mirror. The object is represented as a vertical arrow.

For a convex mirror, these paraxial rays are useful for ray tracing (see Figure 25.21):

Ray 1. This ray leaves the object traveling parallel to the principal axis. After reflection from the mirror, the ray appears to originate from the focal point of the mirror.

Ray 2. This ray leaves the object and heads toward the focal point. After reflection, the ray travels parallel to the principal axis.

Ray 3. This ray leaves the object and travels toward the center of curvature. The ray strikes the mirror perpendicularly and reflects back on itself.

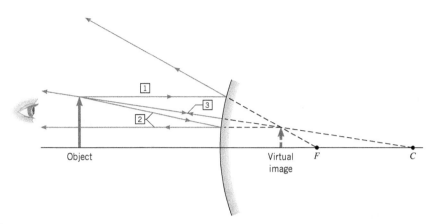

| **FIGURE 25.21**
An object placed in front of a convex mirror always produces a virtual image behind the mirror. The virtual image is reduced in size and upright.

25.6 | The Mirror Equation and the Magnification Equation

The mirror equation specifies the relation between the object distance d_o, the image distance d_i, and the focal length f of the mirror:

$$\frac{1}{d_o} + \frac{1}{d_i} = \frac{1}{f} \tag{25.3}$$

The mirror equation can be used with either concave or convex mirrors.

The magnification m of a mirror is the ratio of the image height h_i to the object height h_o:

$$m = \frac{h_i}{h_o}$$

The magnification is also related to d_i and d_o by the magnification equation:

$$m = -\frac{d_i}{d_o} \tag{25.4}$$

The algebraic sign conventions for the variables appearing in these equations are summarized in the following Reasoning Strategy.

REASONING STRATEGY

Summary of Sign Conventions for Spherical Mirrors

Focal length

f is + for a concave mirror.
f is − for a convex mirror.

Object distance

d_o is + if the object is in front of the mirror (real object).
d_o is − if the object is behind the mirror (virtual object).*

Image distance

d_i is + if the image is in front of the mirror (real image).
d_i is − if the image is behind the mirror (virtual image).

Magnification

m is + for an image that is upright with respect to the object.
m is − for an image that is inverted with respect to the object.

*Sometimes optical systems use two (or more) mirrors, and the image formed by the first mirror serves as the object for the second mirror. Occasionally, such an object falls *behind* the second mirror. In this case the object distance is negative, and the object is said to be a virtual object.

CHAPTER 26

The Refraction of Light: Lenses and Optical Instruments

26.1 | The Index of Refraction

The change in speed as a ray of light goes from one material to another causes the ray to deviate from its incident direction. This change in direction is called refraction. The index of refraction n of a material is the ratio of the speed c of light in a vacuum to the speed v of light in the material:

$$n = \frac{c}{v} \tag{26.1}$$

The values for n are greater than unity, because the speed of light in a material medium is less than it is in a vacuum.

26.2 | Snell's Law and the Refraction of Light

The refraction that occurs at the interface between two materials obeys Snell's law of refraction. This law states that (1) the refracted ray, the incident ray, and the normal to the interface all lie in the same plane, and (2) the angle of refraction θ_2 is related to the angle of incidence θ_1 according to

$$n_1 \sin \theta_1 = n_2 \sin \theta_2 \qquad \text{(26.2)}$$

where n_1 and n_2 are the indices of refraction of the incident and refracting media, respectively. The angles are measured relative to the normal.

Because of refraction, a submerged object has an apparent depth that is different from its actual depth. If the observer is directly above (or below) the object, the apparent depth (or height) d' is related to the actual depth (or height) d according to

$$d' = d\left(\frac{n_2}{n_1}\right) \qquad \text{(26.3)}$$

where n_1 and n_2 are the refractive indices of the materials (the media) in which the object and the observer, respectively, are located.

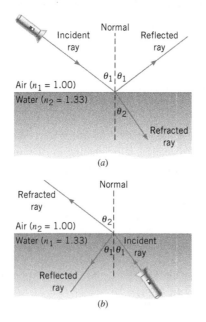

(a)

(b)

| FIGURE 26.1

(a) When a ray of light is directed from air into water, part of the light is reflected at the interface and the remainder is refracted into the water. The refracted ray is bent *toward* the normal ($\theta_2 < \theta_1$).

(b) When a ray of light is directed from water into air, the refracted ray in air is bent *away* from the normal ($\theta_2 > \theta_1$).

26.3 | Total Internal Reflection

When light passes from a material with a larger refractive index n_1 into a material with a smaller refractive index n_2, the refracted ray is bent away from the normal. If the incident ray is at the critical angle θ_c, the angle of refraction is 90°. The critical angle is determined from Snell's law and is given by

$$\sin \theta_c = \frac{n_2}{n_1} \qquad (n_1 > n_2) \tag{26.4}$$

When the angle of incidence exceeds the critical angle, all the incident light is reflected back into the material from which it came, a phenomenon known as total internal reflection.

26.4 | Polarization and the Reflection and Refraction of Light

When light is incident on a nonmetallic surface at the Brewster angle θ_B, the reflected light is completely polarized parallel to the surface. The Brewster angle is given by

$$\tan \theta_B = \frac{n_2}{n_1} \tag{26.5}$$

where n_1 and n_2 are the refractive indices of the incident and refracting media, respectively. When light is incident at the Brewster angle, the reflected and refracted rays are perpendicular to each other.

26.5 | The Dispersion of Light: Prisms and Rainbows

A glass prism can spread a beam of sunlight into a spectrum of colors because the index of refraction of the glass depends on the wavelength of the light. Thus, a prism bends the refracted rays corresponding to different colors by different amounts. The spreading of light into its color components is known as dispersion. The dispersion of light by water droplets in the air leads to the formation of rainbows.

A prism will not bend a light ray at all, neither up nor down, if the surrounding fluid has the same refractive index as the glass, a condition known as index matching.

26.6 | Lenses;
26.7 | The Formation of Images by Lenses

Converging lenses and diverging lenses depend on the phenomenon of refraction in form-ing an image. With a converging lens, paraxial rays that are parallel to the principal axis are focused to a point on the axis by the lens. This point is called the focal point of the lens, and its distance from the lens is the focal length *f*. Paraxial light rays that are parallel to the principal axis of a diverging lens appear to originate from its focal point after pass-ing through the lens. The distance of this point from the lens is the focal length *f*. The image produced by a converging or a diverging lens can be located via a technique known as ray tracing, which utilizes the three rays outlined in the Reasoning Strategy.

REASONING STRATEGY
Ray Tracing for Converging and Diverging Lenses

Converging Lens	Diverging Lens
Ray 1	
This ray initially travels parallel to the principal axis. In passing through a converging lens, the ray is refracted toward the axis and travels through the focal point on the right side of the lens, as Figure 26.25*a* shows.	This ray initially travels parallel to the principal axis. In passing through a diverging lens, the ray is refracted away from the axis, and *appears* to have originated from the focal point on the left of the lens. The dashed line in Figure 26.25*d* represents the apparent path of the ray.
Ray 2	
This ray first passes through the focal point on the left and then is refracted by the lens in such a way that it leaves traveling parallel to the axis, as in Figure 26.25*b*.	This ray leaves the object and moves toward the focal point on the right of the lens. Before reaching the focal point, however, the ray is refracted by the lens so as to exit parallel to the axis. See Figure 26.25*e*, where the dashed line indicates the ray's path in the absence of the lens.
Ray 3*	
This ray travels directly through the center of the thin lens without any appreciable bending, as in Figure 26.25*c*.	This ray travels directly through the center of the thin lens without any appreciable bending, as in Figure 26.25*f*.

*Ray 3 does not bend as it proceeds through the lens because the left and right surfaces of each type of lens are nearly parallel at the center. Thus, in either case, the lens behaves as a transparent slab. The rays incident on and exiting from a slab travel in the same direction with only a lateral displacement. If the lens is sufficiently thin, the displacement is negligibly small.

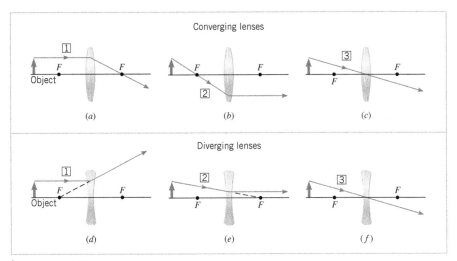

| **FIGURE 26.25**
The rays shown here are useful in determining the nature of the images formed by converging and diverging lenses.

The nature of the image formed by a converging lens depends on where the object is situated relative to the lens. When the object is located at a distance from the lens that is greater than twice the focal length, the image is real, inverted, and smaller than the object. When the object is located at a distance from the lens that is between the focal length and twice the focal length, the image is real, inverted, and larger than the object. When the object is located within the focal length, the image is virtual, upright, and larger than the object.

Regardless of the position of a real object, a diverging lens always produces an image that is virtual, upright, and smaller than the object.

26.8 | The Thin-Lens Equation and the Magnification Equation

The thin-lens equation can be used with either converging or diverging lenses that are thin, and it relates the object distance d_o, the image distance d_i, and the focal length f of the lens:

$$\frac{1}{d_o} + \frac{1}{d_i} = \frac{1}{f} \tag{26.6}$$

The magnification m of a lens is the ratio of the image height h_i to the object height h_o and is also related to d_o and d_i by the magnification equation:

$$m = \frac{h_i}{h_o} = -\frac{d_i}{d_o} \tag{26.7}$$

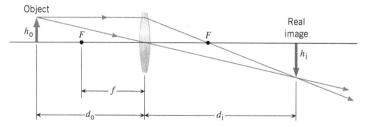

| FIGURE 26.30

The drawing shows the focal length f, the object distance d_o, and the image distance d_i for a converging lens. The object and image heights are, respectively, h_o and h_i.

The algebraic sign conventions for the variables appearing in the thin-lens and magnification equations are summarized in the Reasoning Strategy.

REASONING STRATEGY
Summary of Sign Conventions for Lenses

Focal length

f is $+$ for a converging lens.
f is $-$ for a diverging lens.

Object distance

d_o is $+$ if the object is to the left of the lens (real object), as is usual.
d_o is $-$ if the object is to the right of the lens (virtual object).*

Image distance

d_i is $+$ for an image (real) formed to the right of the lens by a real object.
d_i is $-$ for an image (virtual) formed to the left of the lens by a real object.

Magnification

m is $+$ for an image that is upright with respect to the object.
m is $-$ for an image that is inverted with respect to the object.

*This situation arises in systems containing more than one lens, where the image formed by the first lens becomes the object for the second lens. In such a case, the object of the second lens may lie to the right of that lens, in which event d_o is assigned a negative value and the object is called a virtual object.

26.9 | Lenses in Combination

When two or more lenses are used in combination, the image produced by one lens serves as the object for the next lens.

26.10 | The Human Eye

In the human eye, a real, inverted image is formed on a light-sensitive surface, called the retina. Accommodation is the process by which the focal length of the eye is automatically adjusted, so that objects at different distances produce sharp images on the retina. The near point of the eye is the point nearest the eye at which an object can be placed and still have a sharp image produced on the retina. The far point of the eye is the location of the farthest object on which the fully relaxed eye can focus. For a young and normal eye, the near point is located 25 cm from the eye, and the far point is located at infinity.

A nearsighted (myopic) eye is one that can focus on nearby objects, but not on distant objects. Nearsightedness can be corrected with eyeglasses or contacts made from diverging lenses. A farsighted (hyperopic) eye can see distant objects clearly, but not objects close up. Farsightedness can be corrected with converging lenses.

The refractive power of a lens is measured in diopters and is given by

$$\text{Refractive power (in diopters)} = \frac{1}{f \,(\text{in meters})} \qquad (26.8)$$

where f is the focal length of the lens and must be expressed in meters. A converging lens has a positive refractive power, and a diverging lens has a negative refractive power.

26.11 | Angular Magnification and the Magnifying Glass

The angular size of an object is the angle that it subtends at the eye of the viewer. For small angles, the angular size θ in radians is

$$\theta \,(\text{in radians}) \approx \frac{h_\text{o}}{d_\text{o}}$$

where h_o is the height of the object and d_o is the object distance. The angular magnification M of an optical instrument is the angular size θ' of the final image produced by the instrument divided by the reference angular size θ of the object, which is that seen without the instrument:

$$M = \frac{\theta'}{\theta} \qquad (26.9)$$

A magnifying glass is usually a single converging lens that forms an enlarged, upright, and virtual image of an object placed at or inside the focal point of the lens.

For a magnifying glass held close to the eye, the angular magnification M is approximately

$$M \approx \left(\frac{1}{f} - \frac{1}{d_i}\right)N \qquad (26.10)$$

where f is the focal length of the lens, d_i is the image distance, and N is the distance of the viewer's near point from the eye.

26.12 | The Compound Microscope

A compound microscope usually consists of two lenses, an objective and an eyepiece. The final image is enlarged, inverted, and virtual. The angular magnification M of such a microscope is approximately

$$M \approx -\frac{(L - f_e)N}{f_o f_e} \qquad (L > f_o + f_e) \qquad (26.11)$$

where f_o and f_e are, respectively, the focal lengths of the objective and eyepiece, L is the distance between the two lenses, and N is the distance of the viewer's near point from his or her eye.

26.13 | The Telescope

An astronomical telescope magnifies distant objects with the aid of an objective and an eyepiece, and it produces a final image that is inverted and virtual. The angular magnification M of a telescope is approximately

$$M \approx -\frac{f_o}{f_e} \qquad (26.12)$$

where f_o and f_e are, respectively, the focal lengths of the objective and the eyepiece.

26.14 | Lens Aberrations

Lens aberrations limit the formation of perfectly focused or sharp images by optical instruments. Spherical aberration occurs because rays that pass through the outer edge of a lens with spherical surfaces are not focused at the same point as rays that pass through near the center of the lens.

Chromatic aberration arises because a lens focuses different colors at slightly different points.

CHAPTER 27

Interference and The Wave Nature of Light

27.1 | The Principle of Linear Superposition

The principle of linear superposition states that when two or more waves are present simultaneously in the same region of space, the resultant disturbance is the sum of the disturbances from the individual waves.

Constructive interference occurs at a point when two waves meet there crest-to-crest and trough-to-trough, thus reinforcing each other. When two waves that start out in phase and have traveled some distance meet at a point, constructive interference occurs whenever the travel distances are the same or differ by any integer number of wavelengths: $\ell_2 - \ell_1 = m\lambda$, where ℓ_1 and ℓ_2 are the distances traveled by the waves, and $m = 0, 1, 2, 3, \ldots.$

Destructive interference occurs at a point when two waves meet there crest-to-trough, thus mutually canceling each other. When two waves that start out in phase and have traveled some distance meet at a point, destructive interference occurs whenever the travel distances differ by any odd integer number of half-wavelengths: $\ell_2 - \ell_1 = (m + \frac{1}{2})\lambda$, where ℓ_1 and ℓ_2 are the distances traveled by the waves, and $m = 0, 1, 2, 3, \ldots.$

Two sources are coherent if the waves they emit maintain a constant phase relation. In other words, the waves do not shift relative to one another as time passes. If constructive and destructive interference are to be observed, coherent sources are necessary.

27.2 | Young's Double-Slit Experiment

In Young's double-slit experiment, light passes through a pair of closely spaced narrow slits and produces a pattern of alternating bright and dark fringes on a viewing screen. The fringes arise because of constructive and destructive interference. The angle θ that locates the mth-order bright fringe is given by

$$\sin \theta = \frac{m\lambda}{d} \qquad m = 0, 1, 2, 3, \ldots \qquad (27.1)$$

where λ is the wavelength of the light, and d is the spacing between the slits. The angle that locates the mth dark fringe is given by

$$\sin \theta = \frac{(m + \frac{1}{2})\lambda}{d} \qquad m = 0, 1, 2, 3, \ldots \qquad (27.2)$$

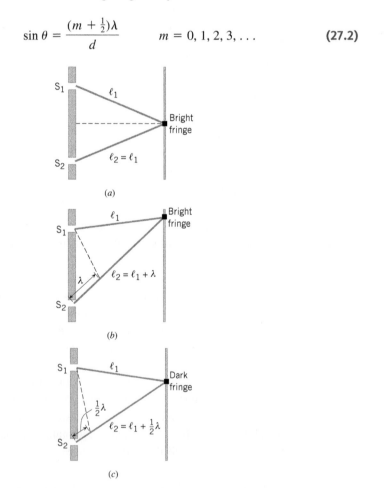

| FIGURE 27.4

The waves that originate from slits S_1 and S_2 interfere constructively (parts *a* and *b*) or destructively (part *c*) on the screen, depending on the difference in distances between the slits and the screen. Note that the slit widths and the distance between the slits have been exaggerated for clarity.

27.3 | Thin-Film Interference

Constructive and destructive interference of light waves can occur with thin films of transparent materials. The interference occurs between light waves that reflect from the top and bottom surfaces of the film.

One important factor in thin-film interference is the thickness of a film relative to the wavelength of the light within the film. The wavelength λ_{film} within a film is

$$\lambda_{\text{film}} = \frac{\lambda_{\text{vacuum}}}{n} \tag{27.3}$$

where λ_{vacuum} is the wavelength in a vacuum, and n is the index of refraction of the film.

A second important factor is the phase change that can occur when light reflects at each surface of the film:

1. When light travels through a material with a smaller index of refraction toward a material with a larger index of refraction, reflection at the boundary occurs along with a phase change that is equivalent to one-half a wavelength in the film.

2. When light travels through a material with a larger index of refraction toward a material with a smaller index of refraction, there is no phase change upon reflection at the boundary.

27.4 | The Michelson Interferometer

An interferometer is an instrument that can be used to measure the wavelength of light by employing interference between two light waves. The Michelson interferometer splits the light into two beams. One beam travels to a fixed mirror, reflects from it, and returns. The other beam travels to a movable mirror, reflects from it, and returns. When the two returning beams are combined, interference is observed, the amount of which depends on the travel distances.

27.5 | Diffraction

Diffraction is the bending of waves around obstacles or around the edges of an opening. Diffraction is an interference effect that can be explained with the aid of Huygens' principle. This principle states that every point on a wave front acts as a source of tiny wavelets that move forward with the same speed as the wave; the wave front at a later instant is the surface that is tangent to the wavelets.

When light passes through a single narrow slit and falls on a viewing screen, a pattern of bright and dark fringes is formed because of the superposition of Huygens wavelets. The angle θ that specifies the mth dark fringe on either side of the central bright fringe is given by

$$\sin\theta = m\frac{\lambda}{W} \qquad m = 1, 2, 3, \ldots \qquad \text{(27.4)}$$

where λ is the wavelength of the light and W is the width of the slit.

27.6 | Resolving Power

The resolving power of an optical instrument is the ability of the instrument to distinguish between two closely spaced objects. Resolving power is limited by the diffraction that occurs when light waves enter an instrument, often through a circular opening.

The Rayleigh criterion specifies that two point objects are just resolved when the first dark fringe in the diffraction pattern of one falls directly on the central bright fringe in the diffraction pattern of the other. According to this specification, the minimum angle (in radians) that two point objects can subtend at a circular aperture of diameter D and still be resolved as separate objects is

$$\theta_{min} \approx 1.22\frac{\lambda}{D} \qquad (\theta_{min} \text{ in radians}) \qquad \text{(27.6)}$$

where λ is the wavelength of the light.

27.7 | The Diffraction Grating

A diffraction grating consists of a large number of parallel, closely spaced slits. When light passes through a diffraction grating and falls on a viewing screen, the light forms a pattern of bright and dark fringes. The bright fringes are referred to as principal maxima and are found at an angle θ such that

$$\sin\theta = m\frac{\lambda}{d} \qquad m = 0, 1, 2, 3, \ldots \qquad \text{(27.7)}$$

where λ is the wavelength of the light and d is the separation between two adjacent slits.

27.8 | Compact Discs, Digital Video Discs, and the Use of Interference

Compact discs and digital video discs depend on interference for their operation.

27.9 | X-Ray Diffraction

A diffraction pattern forms when X-rays are directed onto a crystalline material. The pattern arises because the regularly spaced atoms in a crystal act like a diffraction grating. Because the spacing is extremely small, on the order of 1×10^{-10} m, the wavelength of the electromagnetic waves must also be very small—hence, the use of X-rays. The crystal structure of a material can be determined from its X-ray diffraction pattern.

CHAPTER 28

Special Relativity

28.1 | Events and Inertial Reference Frames

An event is a physical "happening" that occurs at a certain place and time. To record the event an observer uses a reference frame that consists of a coordinate system and a clock. Different observers may use different reference frames.

The theory of special relativity deals with inertial reference frames. An inertial reference frame is one in which Newton's law of inertia is valid. Accelerating reference frames are not inertial reference frames.

28.2 | The Postulates of Special Relativity

The theory of special relativity is based on two postulates. The relativity postulate states that the laws of physics are the same in every inertial reference frame. The speed-of-light postulate says that the speed of light in a vacuum, measured in any inertial reference frame, always has the same value of c, no matter how fast the source of the light and the observer are moving relative to each other.

28.3 | The Relativity of Time: Time Dilation

The proper time interval Δt_0 between two events is the time interval measured by an observer who is at rest relative to the events and views them occurring at the same place. An observer who is in motion with respect to the events and who views them as occurring at different places measures a dilated time interval Δt. The dilated time interval is greater than the proper time interval, according to the time-dilation equation:

$$\Delta t = \frac{\Delta t_0}{\sqrt{1 - \dfrac{v^2}{c^2}}} \qquad (28.1)$$

In this expression, v is the relative speed between the observer who measures Δt_0 and the observer who measures Δt.

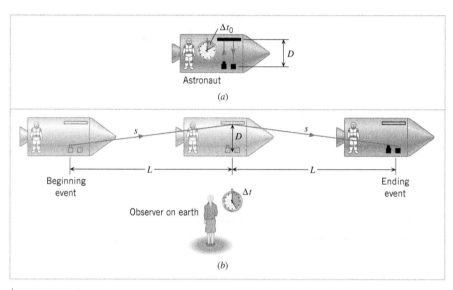

| FIGURE 28.5

(a) The astronaut measures the time interval Δt_0 between successive "ticks" of his light clock.

(b) An observer on earth watches the astronaut's clock and sees the light pulse travel a greater distance between "ticks" than it does in part a. Consequently, the earth-based observer measures a time interval Δt between "ticks" that is greater than Δt_0.

28.4 | The Relativity of Length: Length Contraction

The proper length L_0 between two points is the length measured by an observer who is at rest relative to the points. An observer moving with a relative speed v parallel to the line between the two points does not measure the proper length. Instead, such an observer measures a contracted length L given by the length-contraction formula:

$$L = L_0\sqrt{1 - \frac{v^2}{c^2}} \qquad (28.2)$$

Length contraction occurs only along the direction of the motion. Those dimensions that are perpendicular to the motion are not shortened.

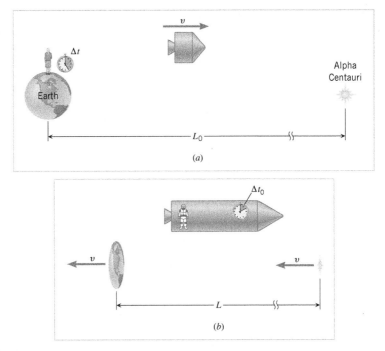

| FIGURE 28.6
(a) As measured by an observer on the earth, the distance to Alpha Centauri is L_0, and the time required to make the trip is Δt.
(b) According to the passenger on the spacecraft, the earth and Alpha Centauri move with speed v relative to the craft. The passenger measures the distance and time of the trip to be L and Δt_0, respectively, both quantities being less than those in part a.

The observer who measures the proper length may not be the observer who measures the proper time interval.

28.5 | Relativistic Momentum

An object of mass m, moving with speed v, has a relativistic momentum whose magnitude p is given by

$$p = \frac{mv}{\sqrt{1 - \dfrac{v^2}{c^2}}} \tag{28.3}$$

28.6 | The Equivalence of Mass and Energy

Energy and mass are equivalent. The total energy E of an object of mass m, moving at speed v, is

$$E = \frac{mc^2}{\sqrt{1 - \dfrac{v^2}{c^2}}} \qquad (28.4)$$

The rest energy E_0 is the total energy of an object at rest ($v = 0$ m/s):

$$E_0 = mc^2 \qquad (28.5)$$

An object's total energy is the sum of its rest energy and its kinetic energy KE, or $E = E_0 + \text{KE}$. Therefore, the kinetic energy is

$$\text{KE} = E - E_0 = mc^2 \left(\frac{1}{\sqrt{1 - \dfrac{v^2}{c^2}}} - 1 \right) \qquad (28.6)$$

The relativistic total energy and momentum are related according to

$$E^2 = p^2 c^2 + m^2 c^4 \qquad (28.7)$$

Objects with mass cannot attain the speed of light c, which is the ultimate speed for such objects.

28.7 | The Relativistic Addition of Velocities

According to special relativity, the velocity-addition formula specifies how the relative velocities of moving objects are related. For objects that move along the same straight line, this formula is

$$v_{AB} = \frac{v_{AC} + v_{CB}}{1 + \dfrac{v_{AC} v_{CB}}{c^2}} \qquad (28.8)$$

where v_{AB} is the velocity of object A relative to object B, v_{AC} is the velocity of object A relative to object C, and v_{CB} is the velocity of object C relative to object B. The velocities can have positive or negative values, depending on whether they are directed along the positive or negative direction. Furthermore, switching the order of the subscripts changes the sign of the velocity, so that, for example, $v_{BA} = -v_{AB}$.

CHAPTER 29

Particles and Waves

29.1 | The Wave–Particle Duality;
29.2 | Blackbody Radiation and Planck's Constant

The wave–particle duality refers to the fact that a wave can exhibit particle-like characteristics and a particle can exhibit wave-like characteristics.

At a constant temperature, a perfect blackbody absorbs and reemits all the electromagnetic radiation that falls on it. Max Planck calculated the emitted radiation intensity per unit wavelength as a function of wavelength. In his theory, Planck assumed that a blackbody consists of atomic oscillators that can have only discrete, or quantized, energies. Planck's quantized energies are given by

$$E = nhf \qquad n = 0, 1, 2, 3, \ldots \qquad (29.1)$$

where h is Planck's constant (6.63×10^{-34} J · s) and f is the vibration frequency of an oscillator.

29.3 | Photons and the Photoelectric Effect

All electromagnetic radiation consists of photons, which are packets of energy. The energy of a photon is

$$E = hf \qquad (29.2)$$

where h is Planck's constant and f is the frequency of the photon. A photon has no mass and always travels at the speed of light c in a vacuum.

The photoelectric effect is the phenomenon in which light shining on a metal surface causes electrons to be ejected from the surface. The work function W_0 of a metal is the minimum work that must be done to eject an electron from the metal. In accordance with the conservation of energy, the electrons ejected from a metal have a maximum kinetic energy KE_{max} that is related to the energy hf of the incident photon and the work function of the metal by

$$hf = KE_{max} + W_0 \qquad (29.3)$$

29.4 | The Momentum of a Photon and the Compton Effect

The magnitude p of a photon's momentum is

$$p = \frac{h}{\lambda} \qquad (29.6)$$

where h is Planck's constant and λ is the wavelength of the photon.

The Compton effect is the scattering of a photon by an electron in a material, the scattered photon having a smaller frequency and, hence, a smaller energy than the incident photon. Part of the photon's energy and momentum are transferred to the recoiling electron. The difference between the wavelength λ' of the scattered photon and the wavelength λ of the incident photon is related to the scattering angle θ by

$$\lambda' - \lambda = \frac{h}{mc}(1 - \cos \theta) \qquad (29.7)$$

where m is the mass of the electron. The quantity $h/(mc)$ is known as the Compton wavelength of the electron.

29.5 | The De Broglie Wavelength and the Wave Nature of Matter

The de Broglie wavelength λ of a particle is

$$\lambda = \frac{h}{p} \qquad (29.8)$$

where h is Planck's constant and p is the magnitude of the relativistic momentum of the particle. Because of its wavelength, a particle can exhibit wave-like characteristics. The wave associated with a particle is a wave of probability.

29.6 | The Heisenberg Uncertainty Principle

The Heisenberg uncertainty principle places limits on our knowledge about the behavior of a particle. The uncertainty principle indicates that

$$(\Delta p_y)(\Delta y) \geq \frac{h}{4\pi} \qquad \text{(29.10)}$$

where Δy is the uncertainty in the particle's position along the y direction, and Δp_y is the uncertainty in the y component of the linear momentum of the particle.

The uncertainty principle also states that

$$(\Delta E)(\Delta t) \geq \frac{h}{4\pi} \qquad \text{(29.11)}$$

where ΔE is the uncertainty in the energy of a particle when it is in a certain state, and Δt is the time interval during which the particle is in the state.

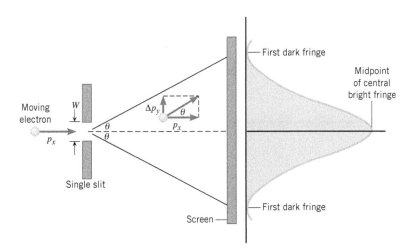

| FIGURE 29.15

When a sufficient number of electrons pass through a single slit and strike the screen, a diffraction pattern of bright and dark fringes emerges. (Only the central bright fringe is shown.) This pattern is due to the wave nature of the electrons and is analogous to that produced by light waves.

CHAPTER 30

The Nature of the Atom

30.1 | Rutherford Scattering and the Nuclear Atom

The idea of a nuclear atom originated in 1911 as a result of experiments by Ernest Rutherford in which α particles were scattered by a thin metal foil. The phrase "nuclear atom" refers to the fact that an atom consists of a small, positively charged nucleus surrounded at relatively large distances by a number of electrons, whose total negative charge equals the positive nuclear charge when the atom is electrically neutral.

30.2 | Line Spectra

A line spectrum is a series of discrete electromagnetic wavelengths emitted by the atoms of a low-pressure gas that is subjected to a sufficiently high potential difference. Certain groups of discrete wavelengths are referred to as "series." The following equations can be used to determine the wavelengths in three of the series that are found in the line spectrum of atomic hydrogen:

$$\textbf{Lyman series} \qquad \frac{1}{\lambda} = R\left(\frac{1}{1^2} - \frac{1}{n^2}\right) \quad n = 2, 3, 4, \ldots \qquad (30.1)$$

$$\textbf{Balmer series} \qquad \frac{1}{\lambda} = R\left(\frac{1}{2^2} - \frac{1}{n^2}\right) \quad n = 3, 4, 5, \ldots \qquad (30.2)$$

$$\textbf{Paschen series} \qquad \frac{1}{\lambda} = R\left(\frac{1}{3^2} - \frac{1}{n^2}\right) \quad n = 4, 5, 6, \ldots \qquad (30.3)$$

The constant term R is called the Rydberg constant and has the value $R = 1.097 \times 10^7 \ \text{m}^{-1}$.

30.3 | The Bohr Model of the Hydrogen Atom

The Bohr model applies to atoms or ions that have only a single electron orbiting a nucleus containing Z protons. This model assumes that the electron exists in circular orbits that are called stationary orbits because the electron does not radiate electromagnetic waves while in them. According to this model, a photon is emitted only when an electron changes from an orbit with a higher energy E_i to an orbit with a lower energy E_f. The orbital energies and the photon frequency f are related according to

$$E_i - E_f = hf \tag{30.4}$$

where h is Planck's constant. The model also assumes that the magnitude L_n of the orbital angular momentum of the electron can only have the following discrete values:

$$L_n = n\frac{h}{2\pi} \qquad n = 1, 2, 3, \ldots \tag{30.8}$$

With these assumptions, it can be shown that the nth Bohr orbit has a radius r_n and is associated with a total energy E_n of

$$r_n = (5.29 \times 10^{-11}\text{m})\frac{n^2}{Z} \qquad n = 1, 2, 3, \ldots \tag{30.10}$$

$$E_n = -(13.6\text{ eV})\frac{Z^2}{n^2} \qquad n = 1, 2, 3, \ldots \tag{30.13}$$

The ionization energy is the minimum energy needed to remove an electron completely from an atom. The Bohr model predicts that the wavelengths comprising the line spectrum emitted by a hydrogen atom can be calculated from

$$\frac{1}{\lambda} = RZ^2\left(\frac{1}{n_f^2} - \frac{1}{n_i^2}\right) \tag{30.14}$$

$$n_i, n_f = 1, 2, 3, \ldots \qquad \text{and} \qquad n_i > n_f$$

30.4 | De Broglie's Explanation of Bohr's Assumption About Angular Momentum

Louis de Broglie proposed that the electron in a circular Bohr orbit should be considered as a particle wave and that standing particle waves around the orbit offer an explanation of the angular momentum assumption in the Bohr model.

30.5 | The Quantum Mechanical Picture of the Hydrogen Atom

Quantum mechanics describes the hydrogen atom in terms of the following four quantum numbers:

1. The principal quantum number n, which can have the integer values $n = 1, 2, 3, \ldots$

2. The orbital quantum number ℓ, which can have the integer values $\ell = 0, 1, 2, \ldots, (n - 1)$

3. The magnetic quantum number m_ℓ, which can have the positive and negative values $m_\ell = -\ell, \ldots, -2, -1, 0, +1, +2, \ldots, +\ell$

4. The spin quantum number m_s, which, for an electron, can be either $m_s = +\frac{1}{2}$ or $m_s = -\frac{1}{2}$

TABLE 30-1

Quantum Numbers for the Hydrogen Atom

Name	Symbol	Allowed Values
Principal quantum number	n	$1, 2, 3, \ldots$
Orbital quantum number	ℓ	$0, 1, 2, \ldots, (n - 1)$
Magnetic quantum number	m_ℓ	$-\ell, \ldots, -2, -1, 0, +1, +2, \ldots, +\ell$
Spin quantum number	m_s	$-\frac{1}{2}, +\frac{1}{2}$

According to quantum mechanics, an electron does not reside in a circular orbit but, rather, has some probability of being found at various distances from the nucleus.

30.6 | The Pauli Exclusion Principle and the Periodic Table of the Elements

The Pauli exclusion principle states that no two electrons in an atom can have the same set of values for the four quantum numbers n, ℓ, m_ℓ, and m_s. This principle determines the way in which the electrons in multiple-electron atoms are distributed into shells (defined by the value of n) and subshells (defined by the value of ℓ).

The following notation is used to refer to the orbital quantum numbers: s denotes $\ell = 0$, p denotes $\ell = 1$, d denotes $\ell = 2$, f denotes $\ell = 3$, g denotes $\ell = 4$, h denotes $\ell = 5$, and so on.

TABLE 30-2

The Convention of Letters Used to Refer to the Orbital Quantum Number

Orbital Quantum Number ℓ	Letter
0	s
1	p
2	d
3	f
4	g
5	h

TABLE 30-3

Ground-State Electronic Configurations of Atoms

Element	Number of Electrons	Configuration of the Electrons
Hydrogen (H)	1	$1s^1$
Helium (He)	2	$1s^2$
Lithium (Li)	3	$1s^2\,2s^1$
Beryllium (Be)	4	$1s^2\,2s^2$
Boron (B)	5	$1s^2\,2s^2\,2p^1$
Carbon (C)	6	$1s^2\,2s^2\,2p^2$
Nitrogen (N)	7	$1s^2\,2s^2\,2p^3$
Oxygen (O)	8	$1s^2\,2s^2\,2p^4$
Fluorine (F)	9	$1s^2\,2s^2\,2p^5$
Neon (Ne)	10	$1s^2\,2s^2\,2p^6$
Sodium (Na)	11	$1s^2\,2s^2\,2p^6\,3s^1$
Magnesium (Mg)	12	$1s^2\,2s^2\,2p^6\,3s^2$
Aluminum (Al)	13	$1s^2\,2s^2\,2p^6\,3s^2\,3p^1$

The arrangement of the periodic table of the elements is related to the Pauli exclusion principle.

30.7 | X-Rays

X-rays are electromagnetic waves emitted when high-energy electrons strike a metal target contained within an evacuated glass tube. The emitted X-ray spectrum of wavelengths consists of sharp "peaks" or "lines," called characteristic X-rays, super-imposed on a broad continuous range of wavelengths called Bremsstrahlung. The K_α characteristic X-ray is emitted when an electron in the $n = 2$ level (L shell) drops into a vacancy in the $n = 1$ level (K shell). Similarly, the K_β characteristic X-ray is emit-ted when an electron in the $n = 3$ level (M shell) drops into a vacancy in the $n = 1$ level (K shell). The minimum wavelength, or cutoff wavelength λ_0, of the Bremsstrahl-ung is determined by the kinetic energy of the electrons striking the target in the X-ray tube, according to

$$\lambda_0 = \frac{hc}{eV} \qquad (30.17)$$

where h is Planck's constant, c is the speed of light in a vacuum, e is the magnitude of the charge on an electron, and V is the potential difference applied across the X-ray tube.

30.8 | The Laser;
30.9 | Medical Applications of the Laser

A laser is a device that generates electromagnetic waves via a process known as stimulated emission. In this process, one photon stimulates the production of another photon by causing an electron in an atom to fall from a higher energy level to a lower energy level. The emitted photon travels in the same direction as the photon causing the stimulation. Because of this mechanism of photon production, the electromag-netic waves generated by a laser are coherent and may be confined to a very narrow beam. Stimulated emission contrasts with the process known as spontaneous emis-sion, in which an electron in an atom also falls from a higher to a lower energy level, but does so spontaneously, in a random direction, without any external provocation.

CHAPTER 31

Nuclear Physics and Radioactivity

31.1 | Nuclear Structure

The nucleus of an atom consists of protons and neutrons, which are collectively referred to as nucleons. A neutron is an electrically neutral particle whose mass is slightly larger than that of the proton. The atomic number Z is the number of protons in the nucleus. The atomic mass number A (or nucleon number) is the total number of protons and neutrons in the nucleus:

$$A = Z + N \tag{31.1}$$

where N is the number of neutrons. For an element whose chemical symbol is X, the symbol for the nucleus is $_Z^A X$.

Nuclei that contain the same number of protons, but a different number of neutrons, are called isotopes.

The approximate radius (in meters) of a nucleus is given by

$$r \approx (1.2 \times 10^{-15}\ \text{m})\, A^{1/3} \tag{31.2}$$

TABLE 31-1

Properties of Select Particles

Particle	Electric Charge (C)	Mass	
		Kilograms (kg)	Atomic Mass Units (u)
Electron	-1.60×10^{-19}	$9.109\,382 \times 10^{-31}$	$5.485\,799 \times 10^{-4}$
Proton	$+1.60 \times 10^{-19}$	$1.672\,622 \times 10^{-27}$	$1.007\,276$
Neutron	0	$1.674\,927 \times 10^{-27}$	$1.008\,665$
Hydrogen atom	0	$1.673\,534 \times 10^{-27}$	$1.007\,825$

31.2 | The Strong Nuclear Force and the Stability of the Nucleus

The strong nuclear force is the force of attraction between nucleons (protons and neutrons) and is one of the three fundamental forces of nature. This force balances the electrostatic force of repulsion between protons and holds the nucleus together. The strong nuclear force has a very short range of action and is almost independent of electric charge.

31.3 | The Mass Defect of the Nucleus and Nuclear Binding Energy

The binding energy of a nucleus is the energy required to separate the nucleus into its constituent protons and neutrons. The binding energy is equal to

$$\text{Binding energy} = (\Delta m)c^2 \qquad \qquad \textbf{(31.3)}$$

where Δm is the mass defect of the nucleus and c is the speed of light in a vacuum. The mass defect is the amount by which the sum of the individual masses of the protons and neutrons exceeds the mass of the intact nucleus.

When specifying nuclear masses, it is customary to use the atomic mass unit (u). One atomic mass unit has a mass of 1.6605×10^{-27} kg and is equivalent to an energy of 931.5 MeV.

31.4 | Radioactivity

Unstable nuclei spontaneously decay by breaking apart or rearranging their internal structure in a process called radioactivity. Naturally occurring radioactivity produces α, β, and γ rays. α rays consist of positively charged particles, each particle being the ^4_2He nucleus of helium. The general form for α decay is

$$^A_Z\text{P} \quad \rightarrow \quad ^{A-4}_{Z-2}\text{D} \quad + \quad ^4_2\text{He}$$

Parent nucleus   Daughter nucleus α particle (helium nucleus)

The most common kind of β ray consists of negatively charged particles, or β^- particles, which are electrons. The general form for β^- decay is

$$^A_Z\text{P} \quad \rightarrow \quad ^A_{Z+1}\text{D} \quad + \quad ^0_{-1}\text{e}$$

Parent nucleus   Daughter nucleus β^- particle (electron)

β^+ decay produces another kind of β ray, which consists of positively charged particles, or β^+ particles. A β^+ particle, also called a positron, has the same mass as an electron, but carries a charge of $+e$ instead of $-e$.

If a radioactive parent nucleus disintegrates into a daughter nucleus that has a different atomic number, as occurs in α and β decay, one element has been converted into another element, the conversion being referred to as a transmutation.

γ rays are high-energy photons emitted by a radioactive nucleus. The general form for γ decay is

$$\underset{\substack{\text{Excited} \\ \text{energy state}}}{{}^{A}_{Z}\text{P*}} \quad \rightarrow \quad \underset{\substack{\text{Lower} \\ \text{energy state}}}{{}^{A}_{Z}\text{P}} \quad + \quad \underset{\substack{\gamma \text{ ray}}}{\gamma}$$

γ decay does not cause a transmutation of one element into another.

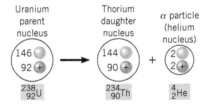

| FIGURE 31.8

α decay occurs when an unstable parent nucleus emits an α particle and in the process is converted into a different, or daughter, nucleus.

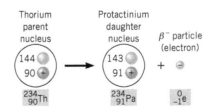

| FIGURE 31.10

β decay occurs when a neutron in an unstable parent nucleus decays into a proton and an electron, the electron being emitted as the β^- particle. In the process, the parent nucleus is transformed into the daughter nucleus.

31.5 | The Neutrino

The neutrino is an electrically neutral particle that is emitted along with β particles and has a mass that is much, much smaller than the mass of an electron.

31.6 | Radioactive Decay and Activity

The half-life of a radioactive isotope is the time required for one-half of the nuclei present to disintegrate or decay. The activity is the number of disintegrations per second that occur. Activity is the magnitude of $\Delta N/\Delta t$, where ΔN is the change in the number N of radioactive nuclei and Δt is the time interval during which the change occurs. In other words, activity is $|\Delta N/\Delta t|$. The SI unit for activity is the becquerel (Bq), one becquerel being one disintegration per second. Activity is sometimes also measured in a unit called the curie (Ci); $1 \text{ Ci} = 3.70 \times 10^{10}$ Bq.

Radioactive decay obeys the following relation

$$\frac{\Delta N}{\Delta t} = -\lambda N \tag{31.4}$$

where λ is the decay constant. This equation can be solved by the methods of integral calculus to show that

$$N = N_0 e^{-\lambda t} \tag{31.5}$$

where N_0 is the original number of nuclei. The decay constant λ is related to the half-life $T_{1/2}$ according to

$$\lambda = \frac{0.693}{T_{1/2}} \tag{31.6}$$

31.7 | Radioactive Dating

If an object contained radioactive nuclei when it was formed, then the decay of these nuclei can be used to determine the age of the object. One way to obtain the age is to relate the present activity A of an object to its initial activity A_0:

$$A = A_0 e^{-\lambda t}$$

where λ is the decay constant and t is the age of the object. For radiocarbon dating that uses the $^{14}_{6}\text{C}$ isotope of carbon, the initial activity is often assumed to be $A_0 = 0.23$ Bq.

31.8 | Radioactive Decay Series

The sequential decay of one nucleus after another is called a radioactive decay series. A decay series starts with a radioactive nucleus and ends with a completely stable nucleus. Figure 31.16 illustrates one such series that begins with uranium $^{238}_{92}U$ and ends with lead $^{206}_{82}Pb$.

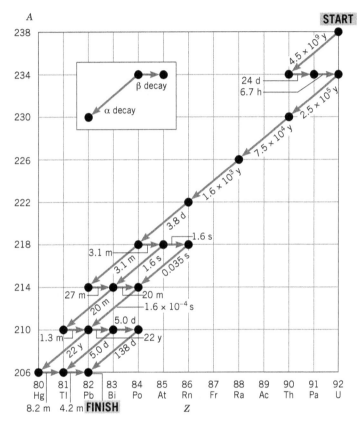

FIGURE 31.16

The radioactive decay series that begins with uranium $^{238}_{92}U$ and ends with lead $^{206}_{82}Pb$. Half-lives are given in seconds (s), minutes (m), hours (h), days (d), or years (y). The inset in the upper left identifies the type of decay that each nucleus undergoes.

31.9 | Radiation Detectors

A number of devices are used to detect α and β particles as well as γ rays. These include the Geiger counter, the scintillation counter, semiconductor detectors, cloud and bubble chambers, and photographic emulsions.

CHAPTER 32

Ionizing Radiation, Nuclear Energy, and Elementary Particles

32.1 | Biological Effects of Ionizing Radiation

Ionizing radiation consists of photons and/or moving particles that have enough energy to ionize an atom or molecule. Exposure is a measure of the ionization produced in air by X-rays or γ rays. When a beam of X-rays or γ rays is sent through a mass m of dry air (0 °C, 1 atm pressure) and produces positive ions whose total charge is q, the exposure in coulombs per kilogram (C/kg) is

$$\text{Exposure (in coulombs per kilogram)} = \frac{q}{m}$$

With q in coulombs (C) and m in kilograms (kg), the exposure in roentgens is

$$\text{Exposure (in roentgens)} = \left(\frac{1}{2.58 \times 10^{-4}}\right)\frac{q}{m} \tag{32.1}$$

The absorbed dose is the amount of energy absorbed from the radiation per unit mass of absorbing material:

$$\text{Absorbed dose} = \frac{\text{Energy absorbed}}{\text{Mass of absorbing material}} \tag{32.2}$$

The SI unit of absorbed dose is the gray (Gy); 1 Gy = 1 J/kg. However, the rad is another unit that is often used: 1 rad = 0.01 Gy.

The amount of biological damage produced by ionizing radiation is different for different types of radiation. The relative biological effectiveness (RBE) is the absorbed dose of 200-keV X-rays required to produce a certain biological effect divided by the dose of a given type of radiation that produces the same biological effect:

$$\text{RBE} = \frac{\begin{array}{c}\text{The dose of 200-keV X-rays that}\\ \text{produces a certain biological effect}\end{array}}{\begin{array}{c}\text{The dose of radiation that}\\ \text{produces the same biological effect}\end{array}} \tag{32.3}$$

The biologically equivalent dose (in rems) is the product of the absorbed dose (in rads) and the RBE:

$$\begin{array}{c}\text{Biologically equivalent dose}\\ \text{(in rems)}\end{array} = \begin{array}{c}\text{Absorbed dose}\\ \text{(in rads)}\end{array} \times \text{RBE} \tag{32.4}$$

32.2 | Induced Nuclear Reactions

An induced nuclear reaction occurs whenever a target nucleus is struck by an incident nucleus, an atomic or subatomic particle, or a γ-ray photon and undergoes a change as a result. An induced nuclear transmutation is a reaction in which the target nucleus is changed into a nucleus of a new element.

All nuclear reactions (induced or spontaneous) obey the conservation laws of physics as they relate to mass/energy, electric charge, linear momentum, angular momentum, and nucleon number.

Nuclear reactions are often written in a shorthand form, such as $^{14}_{7}\text{N}\,(\alpha, p)\,^{17}_{8}\text{O}$. The first and last symbols $^{14}_{7}\text{N}$ and $^{17}_{8}\text{O}$ denote, respectively, the initial and final nuclei.

The symbols within the parentheses denote the incident α particle (on the left) and the small emitted particle or proton p (on the right).

A thermal neutron is one that has a kinetic energy of about 0.04 eV.

32.3 | Nuclear Fission

Nuclear fission occurs when a massive nucleus splits into two less massive fragments. Fission can be induced by the absorption of a thermal neutron. When a massive nucleus fissions, energy is released because the binding energy per nucleon is greater for the fragments than for the original nucleus. Neutrons are also released during nuclear fission. These neutrons can, in turn, induce other nuclei to fission and lead to a process known as a chain reaction. A chain reaction is said to be controlled if each fission event contributes, on average, only one neutron that fissions another nucleus.

32.4 | Nuclear Reactors

A nuclear reactor is a device that generates energy by a controlled chain reaction. Many reactors in use today have the same three basic components: fuel elements, a neutron moderator, and control rods. The fuel elements contain the fissile fuel, and the entire region of fuel elements is known as the reactor core. The neutron moderator is a material (water, for example) that slows down the neutrons released in a fission event to thermal energies so they can initiate additional fission events. Control rods contain material that readily absorbs neutrons without fissioning. They are used to keep the reactor in its normal, or critical, state, in which each fission event leads to one additional fission, no more, no less. The reactor is subcritical when, on average, the neutrons from each fission trigger less than one subsequent fission. The reactor is supercritical when, on average, the neutrons from each fission trigger more than one additional fission.

32.5 | Nuclear Fusion

In a fusion process, two nuclei with smaller masses combine to form a single nucleus with a larger mass. Energy is released by fusion when the binding energy per nucleon is greater for the larger nucleus than for the smaller nuclei. Fusion reactions are said to be thermonuclear because they require extremely high temperatures to proceed. Current studies of nuclear fusion utilize either magnetic confinement or inertial confinement to contain the fusing nuclei at the high temperatures that are necessary.

32.6 | Elementary Particles

Subatomic particles are divided into three families: the boson family (which includes the photon), the lepton family (which includes the electron), and the hadron family (which includes the proton and the neutron).

Elementary particles are the basic building blocks of matter. All members of the boson and lepton families are elementary particles.

The quark theory proposes that the hadrons are not elementary particles but are composed of elementary particles called quarks. Currently, the hundreds of hadrons can be accounted for in terms of six quarks (up, down, strange, charmed, top, and bottom) and their antiquarks.

The standard model consists of two parts: (1) the currently accepted explanation for the strong nuclear force in terms of the quark concept of "color" and (2) the theory of the electroweak interaction.

32.7 | Cosmology

Cosmology is the study of the structure and evolution of the universe. Our universe is expanding. The speed v at which a distant galaxy recedes from the earth is given by Hubble's law:

$$v = Hd \tag{32.5}$$

where $H = 0.022$ m/(s · light-year) is called the Hubble parameter and d is the distance of the galaxy from the earth.

The Big Bang theory postulates that the universe had a definite beginning in a cataclysmic event, sometimes called the primeval fireball. The radiation left over from this event is in the microwave region of the electromagnetic spectrum, and it is consistent with a perfect blackbody radiating at a temperature of 2.7 K, in agreement with theoretical analysis of the Big Bang.

Key Data

Fundamental Constants

Quantity	Symbol	Value*
Avogadro's number	N_A	$6.022\ 141\ 79 \times 10^{23}\ \text{mol}^{-1}$
Boltzmann's constant	k	$1.380\ 6504 \times 10^{-23}\ \text{J/K}$
Electron charge magnitude	e	$1.602\ 176\ 487 \times 10^{-19}\ \text{C}$
Permeability of free space	μ_0	$4\pi \times 10^{-7}\ \text{T}\cdot\text{m/A}$
Permittivity of free space	ϵ_0	$8.854\ 187\ 817 \times 10^{-12}\ \text{C}^2/(\text{N}\cdot\text{m}^2)$
Planck's constant	h	$6.626\ 068\ 96 \times 10^{-34}\ \text{J}\cdot\text{s}$
Mass of electron	m_e	$9.109\ 382\ 15 \times 10^{-31}\ \text{kg}$
Mass of neutron	m_n	$1.674\ 927\ 211 \times 10^{-27}\ \text{kg}$
Mass of proton	m_p	$1.672\ 621\ 637 \times 10^{-27}\ \text{kg}$
Speed of light in vacuum	c	$2.997\ 924\ 58 \times 10^{8}\ \text{m/s}$
Universal gravitational constant	G	$6.674 \times 10^{-11}\ \text{N}\cdot\text{m}^2/\text{kg}^2$
Universal gas constant	R	$8.314\ 472\ \text{J/(mol}\cdot\text{K)}$

* 2006 CODATA recommended values.

Frequently Used Mathematical Symbols

Symbol	Meaning
$=$	is equal to
$\neq$	is not equal to
$\propto$	is proportional to
$>$	is greater than
$<$	is less than
$\approx$	is approximately equal to
$|x|$	absolute value of x (always treated as a positive quantity)
Δ	the difference between two variables (e.g., ΔT is the final temperature minus the initial temperature)
Σ	the sum of two or more variables (e.g., $\sum_{i=1}^{3} x_i = x_1 + x_2 + x_3$)

Useful Physical Data

Acceleration due to earth's gravity	$9.80 \text{ m/s}^2 = 32.2 \text{ ft/s}^2$
Atmospheric pressure at sea level	$1.013 \times 10^5 \text{ Pa} = 14.70 \text{ lb/in.}^2$
Density of air (0 °C, 1 atm pressure)	1.29 kg/m^3
Speed of sound in air (20 °C)	343 m/s
Water	
Density (4 °C)	$1.000 \times 10^3 \text{ kg/m}^3$
Latent heat of fusion	$3.35 \times 10^5 \text{ J/kg}$
Latent heat of vaporization	$2.26 \times 10^6 \text{ J/kg}$
Specific heat capacity	$4186 \text{ J/(kg} \cdot \text{C}°)$
Earth	
Mass	$5.98 \times 10^{24} \text{ kg}$
Radius (equatorial)	$6.38 \times 10^6 \text{ m}$
Mean distance from sun	$1.50 \times 10^{11} \text{ m}$
Moon	
Mass	$7.35 \times 10^{22} \text{ kg}$
Radius (mean)	$1.74 \times 10^6 \text{ m}$
Mean distance from earth	$3.85 \times 10^8 \text{ m}$
Sun	
Mass	$1.99 \times 10^{30} \text{ kg}$
Radius (mean)	$6.96 \times 10^8 \text{ m}$

The Greek Alphabet

Alpha	A	α	Iota	I	ι	Rho	P	ρ
Beta	B	β	Kappa	K	κ	Sigma	Σ	σ
Gamma	Γ	γ	Lambda	Λ	λ	Tau	T	τ
Delta	Δ	δ	Mu	M	μ	Upsilon	Y	υ
Epsilon	E	ϵ	Nu	N	ν	Phi	Φ	ϕ
Zeta	Z	ζ	Xi	Ξ	ξ	Chi	X	χ
Eta	H	η	Omicron	O	o	Psi	Ψ	ψ
Theta	Θ	θ	Pi	Π	π	Omega	Ω	ω

Standard Prefixes Used to Denote Multiples of Ten

Prefix	Symbol	Factor
Tera	T	10^{12}
Giga	G	10^{9}
Mega	M	10^{6}
Kilo	k	10^{3}
Hecto	h	10^{2}
Deka	da	10^{1}
Deci	d	10^{-1}
Centi	c	10^{-2}
Milli	m	10^{-3}
Micro	μ	10^{-6}
Nano	n	10^{-9}
Pico	p	10^{-12}
Femto	f	10^{-15}

Basic Mathematical Formulas

Area of a circle $= \pi r^2$

Circumference of a circle $= 2\pi r$

Surface area of a sphere $= 4\pi r^2$

Volume of a sphere $= \frac{4}{3}\pi r^3$

Pythagorean theorem: $h^2 = h_o^2 + h_a^2$

Sine of an angle: $\sin \theta = h_o/h$

Cosine of an angle: $\cos \theta = h_a/h$

Tangent of an angle: $\tan \theta = h_o/h_a$

Law of cosines: $c^2 = a^2 + b^2 - 2ab \cos \gamma$

Law of sines: $a/\sin \alpha = b/\sin \beta = c/\sin \gamma$

Quadratic formula:

If $ax^2 + bx + c = 0$, then, $x = (-b \pm \sqrt{b^2 - 4ac})/(2a)$

SI Units

Quantity	Name of Unit	Symbol	Expression in Terms of Other SI Units
Length	meter	m	Base unit
Mass	kilogram	kg	Base unit
Time	second	s	Base unit
Electric current	ampere	A	Base unit
Temperature	Kelvin	K	Base unit
Amount of substance	mole	mol	Base unit
Velocity	—	—	m/s
Acceleration	—	—	m/s^2
Force	newton	N	kg·m/s^2
Work, energy	joule	J	N·m
Power	watt	W	J/s
Impulse, momentum	—	—	kg·m/s
Plane angle	radian	rad	m/m
Angular velocity	—	—	rad/s
Angular acceleration	—	—	rad/s^2
Torque	—	—	N·m
Frequency	hertz	Hz	s^{-1}
Density	—	—	kg/m^3
Pressure, stress	pascal	Pa	N/m^2
Viscosity	—	—	Pa·s
Electric charge	coulomb	C	A·s
Electric field	—	—	N/C
Electric potential	volt	V	J/C
Resistance	ohm	Ω	V/A
Capacitance	farad	F	C/V
Inductance	henry	H	V·s/A
Magnetic field	tesla	T	N·s/(C·m)
Magnetic flux	weber	Wb	T·m^2
Specific heat capacity	—	—	J/(kg·K) or J/(kg·C°)
Thermal conductivity	—	—	J/(s·m·K) or J/(s·m·C°)
Entropy	—	—	J/K
Radioactive activity	becquerel	Bq	s^{-1}
Absorbed dose	gray	Gy	J/kg
Exposure	—	—	C/kg

Conversion Factors

Length

1 in. = 2.54 cm

1 ft = 0.3048 m

1 mi = 5280 ft = 1.609 km

1 m = 3.281 ft

1 km = 0.6214 mi

1 angstrom (Å) = 10^{-10} m

Mass

1 slug = 14.59 kg

1 kg = 1000 grams = 6.852×10^{-2} slug

1 atomic mass unit (u) = 1.6605×10^{-27} kg

(1 kg has a weight of 2.205 lb where the acceleration due to gravity is 32.174 ft/s^2)

Time

1 d = 24 h = 1.44×10^3 min = 8.64×10^4 s

1 yr = 365.24 days = 3.156×10^7 s

Speed

1 mi/h = 1.609 km/h = 1.467 ft/s = 0.4470 m/s

1 km/h = 0.6214 mi/h = 0.2778 m/s = 0.9113 ft/s

Force

1 lb = 4.448 N

1 N = 10^5 dynes = 0.2248 lb

Work and Energy

1 J = 0.7376 ft·lb = 10^7 ergs

1 kcal = 4186 J

1 Btu = 1055 J

1 kWh = 3.600×10^6 J

1 eV = 1.602×10^{-19} J

Power

1 hp = 550 ft·lb/s = 745.7 W

1 W = 0.7376 ft·lb/s

Pressure

1 Pa = 1 N/m^2 = 1.450×10^{-4} lb/in.^{2}

1 lb/in.^{2} = 6.895×10^3 Pa

1 atm = 1.013×10^5 Pa = 1.013 bar = 14.70 lb/in.^{2} = 760 torr

Volume

1 liter = 10^{-3} m^3 = 1000 cm^3 = 0.03531 ft^3

1 ft^3 = 0.02832 m^3 = 7.481 U.S. gallons

1 U.S. gallon = 3.785×10^{-3} m^3 = 0.1337 ft^3

Angle

1 radian = 57.30°

1° = 0.01745 radian